GROWING CRYSTALS FROM SOLUTION

VYRASHCHIVANIE KRISTALLOV IZ RASTVOROV

ВЫРАЩИВАНИЕ КРИСТАЛЛОВ ИЗ РАСТВОРОВ

GROWING CRYSTALS FROM SOLUTION

Tomas G. Petrov
Scientific Research Institute for the Earth's Crust
and
Crystallography Section, Leningrad University

and

Evgenii B. Treivus and Aleksei P. Kasatkin
Laboratory for Crystal Growth
Scientific Research Institute for the Earth's Crust
Leningrad University

Translated from Russian by
Albin Tybulewicz
Editor, *Soviet Physics - Semiconductors*

Ⓒ CONSULTANTS BUREAU · NEW YORK · 1969

Tomas Georgievich Petrov, having been graduated in 1954 from Leningrad University with a degree in mineralogy, spent the next two years in postgraduate studies at the Crystallography Section of Leningrad University. This was followed by work as a demonstrator and then a post as a junior scientist. In 1963 he defended a dissertation for the candidate's degree, dealing with the influence of supersaturation and temperature on the genetics of the crystallization of potassium nitrate from aqueous solutions. At present he is a senior scientist at the Scientific Research Institute for the Earth's Crust and teaches at Leningrad University. His scientific interests include the kinetics of crystallization from solutions and its relationship with defects in crystals; the adsorption of impurities, in particular, adsorption of the solvent on the faces of crystals; and the state of a liquid layer at the boundary with a solid.

Evgenii Borisovich Treivus was graduated from Leningrad Mining Institute in 1957. He then did geological work in Siberia. Since 1958 he has been a junior scientist at the Laboratory for Crystal Growth of the Scientific Research Institute for the Earth's Crust at Leningrad University. In 1965 he defended a dissertation for the degree of candidate of geo-mineralogical sciences, dealing with the kinetics and forms of solution of crystals. In addition to research work, Treivus lectures on the growth of crystals in the Geology Department of the University. His scientific interests include the kinetics of crystallization; the formation of inhomogeneities during the growth of crystals; the development of methods for growing crystals from low-temperature solutions; and the modeling of natural crystal-forming processes in low-temperature systems.

Aleksei Prokop'evich Kasatkin was graduated from the Geology Department of Leningrad University in 1960. In 1965 he defended a dissertation for the degree of candidate of geo-mineralogical sciences, dealing with the influence of external conditions on the growth of NaB_2O_3 crystals. Employed next in the Laboratory for Crystal Growth in the Scientific Research Institute for the Earth's Crust as a junior scientist, he was promoted in 1967 to the grade of senior scientist. He lectures on defects in crystals in the Geology Department of the University. His scientific interests include problems of the kinetics of the growth of crystals (influence of the chemical nature of a medium, temperature, supersaturation, fields and radiations, defects, etc.); the development of methods for growing single crystals from low- and high-temperature solutions; and the modeling of natural crystal-forming processes in low-temperature systems.

First Printing – May 1969
Second Printing – March 1971

The original Russian text, published by Nedra Press in Leningrad in 1967, has been corrected by the authors for the present edition.

Томас Георгиевич Петров
Евгений Борисович Трейвус
Алексей Прокопьевич Касаткин

ВЫРАЩИВАНИЕ КРИСТАЛЛОВ ИЗ РАСТВОРОВ

Library of Congress Catalog Card Number 69-12536
SBN 306-10824-0

© 1969 Consultants Bureau
A Division of Plenum Publishing Corporation
227 West 17 Street, New York, N. Y. 10011
All rights reserved

No part of this publication may be reproduced in any
form without written permission from the publisher
Printed in the United States of America

PREFACE TO THE AMERICAN EDITION

The scientific aspects of this monograph are described in the Preface to the Russian edition; here, we would like to describe how we wrote it.

We found that the repeated description and presentation of the same material in lectures was boring. We were granted a leave of absence by our Institute, took a tent and left the city. We caught fish in a forest lake covered with water-lilies, played ball, and made motion pictures. After several days, it started to rain, but by this time the rough draft of the book was ready. We continued to add, delete, alter, etc. Much time and effort went into the writing of this book and we hope that our life and that of others interested in crystal growth will be made easier by the publication of this monograph.

Naturally, we are very pleased that our book will be read abroad and we thank Plenum Publishing Corporation for publishing it in English.

May 12, 1968

T. G. Petrov
E. B. Treivus
A. P. Kasatkin

PREFACE TO THE RUSSIAN EDITION

The growth of crystals has been of interest to geologists for a long time because this process occurs widely in nature. Relatively recently, the growth of crystals has become a rapidly developing branch of chemical technology. Single crystals are nowadays used in various branches of technology because of their many and unique physical properties. However, if one wishes to become acquainted with the practice of growing of crystals one finds little published material, particularly in the case of growth from solutions. The problems which arise can be divided into two groups:

a) the selection of an appropriate growth method, the construction of crystallization apparatus, and the operational procedure;

b) the measures which need to be taken to prevent inhomogeneities from occurring in crystals and other anomalies.

As far as the authors are aware, the growth of crystals has not yet been dealt with from this point of view. Descriptions of various types of crystallization apparatus and of methods of growth are given in the books by A. V. Shubnikov (1935), H. E. Buckley (1951), and V. D. Kuznetsov (1953). These monographs also contain extensive practical material on the characteristic features of the growth of crystals but none of them can be used as a handbook. They have very few descriptions of growing procedures. All three are badly out of date. Moreover, Shubnikov's book has become a bibliographical rarity. A book by O. M. Ansheles, V. B. Tatarskii, and A A. Shternberg (1945) has partially served as our guide. However, this work is limited to the description of only one growing method, deals only with one substance, and it is difficult to find this monograph in bookshops. Useful books have been published in English (A. Holden and P. Singer, 1960; A. Van Hook, 1961) and in German (K. T. Wilke, 1963) but these again do not fulfill the purpose we had in mind.

The present monograph is intended to be a guide to the growth of crystals from solutions at atmospheric pressure and moderate temperatures (up to 100°C). The growth of crystals from solutions under these conditions is quite easy in the laboratory. The preparation of crystals from solutions is and will remain an important method in the production of various crystals. It is important in the preparation of materials which melt incongruently (they decompose at the melting point) and in the preparation of those polymorphic modifications which do not crystallize from the melt.

The methods described in this book are intended mainly for the growth of small crystals, ranging from a fraction of a millimeter to 1-2 cm, which are needed to investigate the habit of such crystals and their growth features, properties, structure, etc.

The authors leaned heavily on their own experience. They were careful not to describe published methods without trying them out themselves. This resulted in some bias in the decriptions of details of methods and procedures. Many well-known special variants of the growth of crystals of some substances are not given in the present book. The authors were of the opinion that the chief aim was to get the reader to understand the basis of the mechanism of crystal growth and to learn a certain number of well-established experimental procedures. A thoughtful reader will always find new variants of a method and it is easier to develop such variants independently than to find them in the literature.

The book is intended for a wide range of readers. It is meant for physicists and chemists interested in the growth of small crystals required for laboratory investigations. It is also intended for experimental geologists who are interested in the processes of crystal growth. It may be used by students in the practical work which is carried out in some higher educational establishments. We hope that the book will be helpful also to members of the staff of specialist crystallization laboratories and to workers in industry.

The book covers a narrower range of topics than that implied by its title. However, the title is not accidental. The material presented is also of interest to specialists concerned with crystal growth under other conditions (from molten solutions, by hydrothermal methods, from gases, and from melts) because the general principles of growth are the same under different conditions.

The book was much improved by discussions of the manuscript, especially of some of its parts and of specific problems, with Professor V. B. Tatarskii, Professor I. I. Shafranovskii, and engineers V. K Andriashin, A. E. Glikin, A. N. Kovalevskii, Yu. O. Punin, and T. P. Ul'yanova. The authors are very grateful for this help. They are also grateful to Ch. S. Soboleev for supplying the photographs of growth spirals on crystals.

Readers' comments will be welcome at the following address: Leningrad F-2, ul. Lomonosova 22, Leningradskoe Otdelenie Izdatel'stva "Nedra" (Leningrad Division of "Nedra" Press).

CONTENTS

CHAPTER I. FUNDAMENTALS OF THE THEORY OF GROWTH OF CRYSTALS FROM SOLUTIONS 1
   §1. Binding of Particles and Phase State 1
   §2. Structure of Real Crystals 3
   §3. Crystallization Medium 8
   §4. Crystal Growth Mechanisms 14
   §5. Role of Strongly Adsorbed Impurities 20
   §6. Volume Diffusion during Growth of Crystals 26

CHAPTER II. METHODS FOR GROWING CRYSTALS 31
   §7. Classification of Methods for Growing Crystals 31

   A. CRYSTALLIZATION UNDER STEADY-STATE CONDITIONS 33
   §8. Crystallization under Thermal Convection Conditions 33
   §9. Crystallization under Concentration Convection Conditions 37
   §10. Crystallization under Forced Convection Conditions 40
   §11. Crystallization by Chemical Reactions under Counter-Diffusion Conditions 44

   B. CRYSTALLIZATION UNDER NON-STEADY-STATE CONDITIONS 48
   §12. Crystallization by Cooling of Solutions 48
   §13. Crystallization by Solvent Evaporation 52

CHAPTER III. TECHNICAL EQUIPMENT FOR A CRYSTAL-GROWING LABORATORY 56
   §14. Laboratory Buildings and General Equipment 56
   §15. Thermostats 57
   §16. Devices for Producing Relative Motion in a Crystal–Solution System . 61
   §17. Filtration Devices and Methods 65
   §18. Materials Used in Crystal-Growing Apparatus 66

CHAPTER IV. PREPARATION FOR AND CONTROL OF GROWTH OF CRYSTALS 69
   §19. Acquisition of Information on a Substance and Selection of a Solvent . . 69
   §20. Preliminary Information on Growth of Crystals 71
   §21. Some Methods for Control of Growth Conditions and Crystal Quality . . 73
   §22. Selection of Crystal Growth Method 77

CHAPTER V. EXPERIMENTAL PROCEDURES ... 82
§23. Purification of Reagents ... 82
§24. Preparation of Solutions ... 84
§25. Determination of Solubility ... 86
§26. Determination of Saturation Temperatures of Solutions ... 87
§27. Preparation of Seed Crystals ... 91
§28. Crystal Holders and Seed-Mounting Methods ... 92
§29. How to Treat a Grown Crystal ... 95
§30. Identification of Crystals ... 96

CONCLUSIONS ... 98

LITERATURE CITED ... 99

CHAPTER I

FUNDAMENTALS OF THE THEORY OF GROWTH OF CRYSTALS FROM SOLUTIONS

§1. Binding of Particles and Phase State*

All chemical bonds are due to the tendency of the potential energy of particles to decrease as the particles approach each other, because of the interaction of their outer electron shells. This reduction in the energy is due to the overlap of the electron shells of the atoms or due to the electrostatic (Coulomb) attraction. However, beginning from a certain distance between the centers of the atoms, the repulsion forces begin to predominate and the energy of the particles increases rapidly. In general, the potential energy curve (the energy of interaction between two atoms) varies with the distance between the two atoms as shown in Fig. 1. R_0 represents the most stable (equilibrium) distance between two nuclei or the interatomic distance. At all temperatures (including 0°K), an atom which has a finite kinetic energy vibrates near this equilibrium position.

The binding energy is the energy liberated as a result of the interaction between particles. The energies for different types of binding are listed in Table 1.

In the presence of hydrogen and van der Waals bonds in a crystal, we can still identify separate molecules. However, if a crystal consists of particles between which metallic, covalent or ionic bonds are formed, we cannot identify a single molecule. In such substances, the whole crystal acts as a molecule. Molecules consisting of the minimum number of atoms prescribed by the chemical formula (for example, NaCl) can be obtained either by dissolving a crystal in a suitable solvent in which such a molecult does not dissociate, or by evaporation. Thus, in the vapor state alkali halides consist of diatomic molecules.

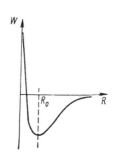

Fig. 1. Dependence of the potential energy of interaction between two atoms on the distance between them.

All the interatomic and intermolecular forces decrease rapidly when the distance between the interacting particles is increased. On the other hand, an increase in the kinetic energy of interacting particles to a value comparable with the binding energy between them may result in the breaking of bonds. These two factors account for the effects of temperature and pressure on matter. An increase in the temperature (an increase in the vibrational energy) sooner or later results in the breaking of the bonds be-

---

*The nature of bonds between particles (atoms, ions, molecules) is discussed in detail in many general text books on chemistry and physics, particularly in the works of Kondrat'ev (1959) and Spice (1964), in appropriate articles in the Russian Encylopedic Physics Dictionary (1962-1966), and elsewhere.

TABLE 1

Type of bond	Molecules	Binding energy	Molecules	Binding energy
Covalent	N_2	946.8	SiC	435.8
	C_2	604.2	HCl	428.2
	O_2	499.1	Br_2	193.1
Ionic	LiF	579.5	NaCl	412.7
	KCl	426.5	NaI	300
Metallic	Au_2	219.6	NaK	63.7
	Al_2	192.7	Mg_2	33.5
	Pb_2	100.6		
Hydrogen	$CH_3COOH-CH_3COOH$	34.4		
	NH_3-NH_3	5.4	H_2O-H_2O	18.9
van der Waals	Kr−Kr	1.40		
	Ar−Ar	0.96	Ne−Ne	0.32

Note. The results for the covalent, ionic, and metallic bonds are taken from a book by Vedeneev et al. (1962) and they apply at 25°C. The data on the hydrogen bond are taken from the Russian Encyclopedic Physics Dictionary (vol. 1, p. 290). The temperatures for the hydrogen bond examples are not given in this Dictionary. The data on the van der Waals bond are taken from Kondrat'ev's book (1959, p. 445). Againt, the temperature is not given. In all these cases, the binding energies have been converted from kcal/mole into kJ/mole.

TABLE 2

Property	Gas	Condensed state	
		liquid	crystal
Order in particle distribution	None	Short-range but no long-range order	Long-range order
Distance between centers of particles R compared with particle size r	$R \gg r$	$R \approx r$	$R \approx r$
Nature of main type of motion	Translation	Vibration and translation	Vibration
Strength of bonds between particles	No bonds	Increasing strength of bonds from liquids to crystals	

tween the particles and gradually transforms a substance from the solid state to the liquid and then to the gaseous state, or directly from the solid to the gaseous state. An increase in the external pressure produces effects which are, in general, opposite to those produced by an increase in temperature. By reducing the distance between particles, the pressure tends to form chemical bonds. If a crystal has a loosely packed structure, an increase in the pressure may also break the bonds and produce melting. However, a continued increase in the pressure alters the nature of the chemical bonds and produces crystals with metallic binding and close-packed structures, which are not typical of these compounds at atmospheric pressure. In general, an increase of the pressure in a gas or a liquid increases the probability of collisions between particles and therefore the probability of the formation of bonds. The lifetime of such bonds is governed by the relationship between the average kinetic energy of the particles, i.e., the temperature and the potential energy, which represents the strength of the bond.

Thus, depending on the pressure, the temperature, and the strength of the chemical bonds, a substance may exist in three main states: crystalline, liquid, and gaseous. Table 2 lists the comparable microscopic properties of these states.

There are two more states, which can be regarded as intermediate between liquids and crystals: they are the glassy and amorphous states. Recent investigations have established the presence of short-range order and the absence of long-range order in the distribution of particles in the glassy and amorphous (resins, plastics) states. The absence of appreciable differences between these two states allows us to regard them as one state, which is similar to liquids but in which the mobility of particles is much less than in liquids.

§2. Structure of Real Crystals

An ideal crystal can be defined as an infinitely large system consisting of a regular space lattice at each point of which there are identical particles or identical groups of particles. These particles in a crystal may be atoms, ions, or molecules.

When a real crystal is formed many factors prevent the formation of a perfectly regular lattice of particles, and the crystal may capture foreign particles. Thus, departures from the ideal structure, known as defects, appear in a real crystal.

A real crystal is formed as a result of two opposing tendencies: (1) a tendency to achieve an ordered distribution of particles so as to obtain a maximum compensation of chemical bonds; ordering is realized most easily when the particles are spherical and the number of types of particle in a crystal is strictly limited; (2) a tendency to mixing and disorder due to the thermal motion of particles.

These tendencies depend on the actual conditions (the chemical nature of the medium, temperature, etc.) under which crystals are grown. It follows that crystals of different degrees of perfection can be obtained. These two tendencies obey a general law of a decrease in the free energy in a system in any process.*

Textbooks on crystallography usually describe in detail ideal crystals (or, more exactly, their models) but pay little attention to the disorder. Therefore, we shall consider defects in crystals, and refer the reader to crystallography textbooks [Popov and Shafranovski (1964); Ansheles (1952)] for information on ideal crystals.

Defects can be classified conveniently according to their size. If we assume that the structure units in a crystal (atoms, ions, and molecules) have zero dimensions, we can distinguish the following types of defect: (1) zero-dimensional; (2) one-dimensional; (3) two-dimensional; and (4) three-dimensional.

1. Zero-Dimensional or Point Defects.
The main types of such defect are vacancies (unfilled sites in a crystal), interstitial atoms, and any other impurity particles [van Bueren (1961); Frank-Kamenetskii (1964)]. Vacancies and interstitial atoms are generated primarily by the thermal motion of atoms. The higher the temperature, the larger is the number of vacancies and interstitials in a crystal. Moreover, vacancies are formed when the valence of impurities introduced into a crystal differs from the valence of the particles in the host

---

*An elementary account of thermodynamics is given in Veinik's book (1965).

matrix. Thus, for example, the presence of Ca^{2+} impurities in NaCl crystals produces cation vacancies in amounts corresponding to the impurity concentration. Moreover, all minerals contain linearly oriented arrays of point defects along tracks of nuclear fragments of radioactive elements [Sviridov (1964)].

Point defects in crystals are, in general, distributed nonuniformly. The cause of this will be discussed in §5.

2. One-Dimensional (Line) Defects: Dislocations [Read (1953); van Bueren (1961)]. By specifying certain rules, we can draw a closed contour through identical points in an ideal crystal structure (for example, we can draw this contour through the lattice sites). This contour is known as the Burgers circuit. The number of steps in such a circuit is equal in the positive and the negative directions along each axis. Usually, the Burgers circuit is drawn in one plane in a crystal.

A real crystal does not have an ideal structure but various parts of its are disturbed to some extent. We shall define as a good material a substance in which particles suffer only elastic deformations. In a good material, the Burgers circuit around a zero-dimensional defect is closed. However, if a contour drawn around a defect is open (Fig. 2), such a defect is one-dimensional and is called a dislocation. It is evident from Fig. 2 that in this case a crystal has an extra half-plane in the vertical direction. The edge of this half-plane is known as the dislocation axis.

The gap separating the ends of the contour around the dislocation is known as the modulus (absolute value) of the Burgers vector b. Thus, a closed contour around a dislocation is equal to the sum of the Burgers circuit and the Burgers vector. Two extreme cases of the position of b with respect to the dislocation axis can be distinguished: correspondingly, there are two main types of dislocation. When b is perpendicular to the dislocation axis, we have an edge dislocation (Fig. 2). Naturally, more than one half-plane may be introduced. The Burgers vector is then correspondingly larger. If b is parallel to the dislocation axis we have a screw dislocation. The nature of a screw dislocation can be seen from Fig. 3. Edge and screw dislocations are the extreme forms of dislocations. There are also dislocations with b having an intermediate orientation with respect to the dislocation axis. They are known as mixed dislocations. There are also other types of dislocation.

We shall mention some important properties common to all dislocations.

The Burgers vector along a dislocation is constant and therefore a dislocation cannot end in a crystal.

The energy of a dislocation, due to the deformation of a crystal around the dislocation core, depends on the mechanical properties of the crystal and is directly proportional to the square of the Burgers vector. This is why the existence of dislocations with large (much larger than the parameters of the unit cell in a crystal) Burgers vectors is not favored by energy considerations. Therefore, a dislocation with a large value of b usually splits (during its growth) into a group of dislocations the sum of whose Burgers vectors is equal to the vector of the original dislocation but whose total energy is less. This explains the following observation. Cubic crystals of sodium bromate, grown from aqueous solutions, sometimes have a zonal structure indicated by an anomalous birefringence. Using a microscope giving sufficient magnification, we can see that the birefringence is concentrated in zones consisting of lines aligned along the growth axis. These lines end at the boundary of a zone. Obviously, dislocations with large Burgers vectors formed in these crystals during growth give rise to a considerable birefringence, but this birefringence vanishes when such dislocations split into smaller ones.

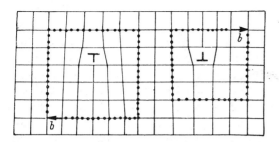

Fig. 2. Burgers circuits (represented by dots), closed by Burgers vectors b, around two edge dislocations of opposite signs. The symbol ⊥ represents the edge of a half-plane.

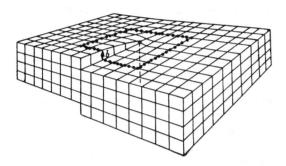

Fig. 3. Screw dislocation with its axis (pointing into the interior of a crystal) shown dashed.

The interaction between the elastic-force fields of dislocations with parallel Burgers vectors results in a repulsion between the dislocations. Dislocations with oppositely directed Burgers vectors attract each other. Dislocations may move under the action of any external and internal stresses in a crystal and they can move due to the diffusion of point defects.

Dislocations in a crystal form a more or less complex dislocation network.

The spatial orientation of dislocations depends on their nature and on the structure of the crystal, primarily on the strength of bonds along various directions.

The concentration of dislocations, defined as the number of dislocations intersecting 1 cm^2 of a given surface, is usually of the order of $10-10^5$ cm^{-2}. Deformed crystals contain 10^6-10^{12} cm^{-2} dislocations. When the temperature is increased, the dislocations spread more or less rapidly and the number of dislocations decreases. This is used in practice to relieve stresses, and is known as annealing. At high impurity concentrations impurities can pin down dislocations and reduce their mobility. The concentration of impurities near a dislocation core is usually higher than elsewhere. Such an accumulation of impurities is frequently called a Cottrell cloud.

Crystals exhibit a higher chemical activity near dislocation cores. Therefore, melting, oxidation, and dissolution always begin at defects such as dislocations, and are more rapid near defects. This property is used in the most convenient method for the visualization of defects by etching, i.e., a slow dissolution of a crystal. The shape of the resultant etch pits can be used to deduce the nature of the defects responsible for these pits. Thus, two types of etch pits can always be distinguished on the surface of a crystal: a) pits with a sharp bottom, i.e., pyramidal pits whose depth increases during the dissolution of the crystal, and which appear again in the original places after polishing (these pits are the points of emergence of screw and edge dislocations; a shift of the apex of the pyramid with respect to the center of the pit indicates a deviation of the dislocation axis from the normal to the surface); b) pits with a flat bottom in the form of a truncated pyramid (these pits exist only briefly and they disappear rapidly during dissolution; they represent accumulations of point defects).*

We can mention here that the points of emergence of screw dislocations at the surface of a crystal play an important role in its growth. Groups of dislocations and single dislocations may act as growth centers on a face.

---

*The selection of etchants as well as other methods for the visualization of dislocations are reviewed in the bibliographical guides [cf. Dislocations in Crystals (1960, 1966)].

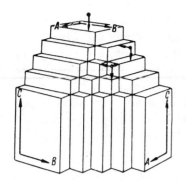

Fig. 4. Types of crystal face classified in accordance with surface bonds (A, B, C represent the vectors of strong bonds). Points with arrows represent atoms adsorbed, respectively, on an F face (single arrow), on an S face (two arrows), and on a K face (three arrows).

3. Two-Dimensional Defects. These defects include, first of all, the actual surfaces of a crystal. The designation of a surface as a defect does not have just a formal meaning in the sense of a departure from periodicity in the distribution of structure units. The surface has an excess energy compared with the energy of the particles in the interior of the crystal; this is known as the surface energy. This energy is due to an irregular distribution of particles near the surface, compared with the regular distribution in the interior. In particular, de Boer (1959) has investigated the faces of cubic alkali halides and has found that a plane passing through the centers of anions is ~0.2 Å higher than a plane passing through the centers of cations. In the interior, these planes coincide. The anion-cation distances in gas molecules [Kondrat'ev (1959), p. 277] are 20% smaller than the corresponding distances in crystals. The cause of this phenomenon is the unsaturated nature of ionic bonds, which results in a strong dependence of the bond length on the number of nearest neighbors.

Since the surface charge of an ionic crystal is not completely compensated, we can expect tensile forces along the surface of the crystals which may be strong enough to produce microcracks.

Because of the lattice structure of crystals, the atomic structure of different faces is different. The characteristic features of the atomic structure of the faces can be found by projecting the structure of the crystal as a whole onto the plane of the face. Various methods of plotting such projections have been described by Treivus and Petrov (1964) and by Evzikova (1965). Detailed reviews are available, which summarize the results of investigations of the atomic structure of crystal faces and the various methods of classification [Honigmann (1958)]. All surfaces are usually divided into three types (Fig. 4):

a) smooth surfaces (type F): along these surfaces there are at least two series of nonparallel chains of strong bonds;

b) stepped surfaces (type S): there is at least one series of chains of strong bonds parallel to such surfaces;

c) "cellular" surfaces (type K): there are no series of chains of strong bonds along these surfaces.

We can easily see that the strength of the bonds of particles adsorbed by these surfaces increases in the same order. Consequently, the rate of growth of these faces increases in the same order. The habit of a crystal (its faces) is governed by purely geometrical relationships between the rates of growth of the various faces. The rapidly growing faces usually become smaller and therefore the habit of a crystal is governed by the slowly growing faces. These slowly growing faces usually belong to the first two types of surface. Usually, only a relatively small number of crystallographic faces may exist in a given crystal and these faces are characterized by rational parameters [Ansheles (1952); Popov and Shafranovskii (1964)].

When a crystal is examined under a microscope, it is difficult to find a face which is perfectly smooth. Usually, the surface of a face of a growing crystal has stripes, projections of various shapes, and sometimes a very complex topography which only partly obeys the ideal symmetry rules. The tips of the projections are at the points of emergence of dislocations.

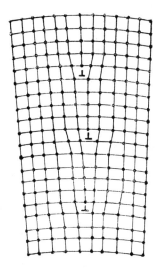

Fig. 5. Dislocation boundary between two blocks in a crystal.

This is confirmed by etching the octahedral faces of potash alum and the tetrahedral faces of sodium bromate.

The macroscopic features of the surfaces of crystals (their topography) have been described in detail by Shafranovskii (1957; 1960; 1961). The morphology of crystal faces provides a more sensitive indication of the crystallization conditions than the habit and, in principle, the morphology represents the crystallization conditions in the latest stage of formation of a crystal. The nature of crystal faces cannot, in general, be used to deduce the conditions of formation of the interior of a crystal.

A boundary between slightly misaligned blocks in one crystallite, or a boundary between two crystallites is also a two-dimensional defect (Fig. 5). Such two-dimensional defects are arrays of parallel dislocations and they can be made visible by etching, which produces chains of etch pits.

Sometimes a more or less flat surface has faceted projections representing an almost parallel combination of a possibly large number of crystallites (Fig. 47). Many external single-crystal features are found to be nearly parallel combinations of intergrown crystallites. In principle, we cannot distinguish with certainty between single crystals and intergrowths, particularly in the case of crystals prepared from melts. In the case of crystals grown from solutions, we can regard as single crystals those crystals which may contain blocks but which have no optically observable internal boundaries. The splitting of crystals consisting of blocks does not usually take place along block boundaries whereas intergrown crystals are separated by what are known as induced faces. These faces are always stepped and there are inclusions of the medium from which an intergrown projection has originated. The cleaving of such intergrown crystallites takes place mainly along the boundaries of the individual crystallites.

Twin boundaries are another form of two-dimensional defect. Twins may consist of several crystals (between two and four) or a large number of crystals (examples of twins are shown in Figs. 50 and 51).

4. **Three-Dimensional Defects**. These include primarily closed cavities (inclusions) in a crystal. Depending on the phase state of the ambient medium during crystallization, they may be filled with a vapor, a liquid, or with solid foreign particles, or they may consist of several phases. When a crystal containing liquid or gaseous inclusions is kept for a long time at a temperature ensuring that the concentration of the host-crystal particles within the cavities increases appreciably, the usually shapeless inclusions are replaced by faceted ones, which are known as "negative crystals." Small solution-containing inclusions in crystals exhibiting a high solubility at room temperature become faceted in several days.

The content of inclusions in mineral crystals gives extensive information on the physicochemical state of the medium at various stages of the lifetime of such crystals [Ermakov (1950)].

Three-dimensional defects also include open cracks in crystals. Healed cracks represent combinations of three-dimensional and two-dimensional defects. The healing of a crack consists of the joining of opposite walls. Usually, a crack captures foreign substances (a solution or air) so that the healing is almost always accompanied by the formation of a network of microscopic inclusions along the former surface of a crack. In those cases where the walls close up completely, we observe a two-dimensional system of dislocations. Therefore, an array of emergent dislocations can be regarded as a trace of a healed crack, and it can show whether a given crack has emerged on an external surface or not.

In some cases, complex crystals formed at high temperatures are found to be unstable at low temperatures. They usually split into two structures. One of the crystal phases may form inclusions within the main solid phase. Precipitates of the second phase will thus be three-dimensional defects in the crystal lattice of the main phase. We must mention that diffusion in the solid state, which accompanies the process of precipitation, is affected by the zero-dimensional and one-dimensional defects of the crystal structure. Sometimes, particularly in naturally occurring mica, we find that the transformation of muscovite to biotite takes place along dislocation axes. This can be easily understood. The distorted environment of atoms near dislocation axes makes the atoms along these axes more mobile and more reactive.

Various types of defect are closely related. Thus, in the presence of dislocations, point defects appear more easily and the three-dimensional defects (inclusions formed during growth) are usually accompanied by dislocations [Khadzhi (1966)], while local accumulations of point defects frequently give rise to three- and two-dimensional defects in the form of cracks and dislocations [Shternberg (1962)].

§3. Crystallization Medium

The chemical nature and structure of a crystallization medium govern the habit of a crystal and its perfection within the limits of a possible actual crystal structure.

We shall not consider gaseous and solid crystallization media, but concentrate solely on liquids. In the simplest case, a liquid may be obtained by the congruent melting (melting without decomposition) of a pure compound. Such a liquid is known as a melt. The composition of the crystals growing from such a liquid is the same as the composition of the liquid itself. The crystallization temperature is constant and has the maximum value possible in the crystallization of this compound (at a given pressure). It is equal to the melting point of this compound or is close to it.

If the liquid is a macroscopically homogeneous mixture of different substances, the temperatures of crystallization of the individual components are lower and such a mixture is a solution. Crystals are, in most cases, prepared from solutions containing two substances; these are known as binary solutions. One of the substances is the solvent and the other the solute. Moreover, a solution contains small amounts of other substances, which are regarded as impurities. A perfectly pure substance is as difficult to prepare as a perfectly ordered crystal. In some cases impurities are introduced into a solution deliberately because of their effect on the growth of crystals.

Solutions are usually described in terms of the concentrations of substances present in them.

Let us consider a typical solubility diagram (Fig. 6). The whole concentration-temperature field is separated by the saturated-solution line (solubility curve) into two regions: unsaturated and supersaturated solutions. Saturated solutions are those mixtures which can retain their equilibrium indefinitely in contact with the solid phase with respect to which they are saturated. The solubility of most substances increases when the temperature is increased (the temperature coefficient of the solubility is positive). Crystals can be dissolved only in unsaturated solutions. They crystallize only from supersaturated solutions which contain an excess of the solute above the equilibrium value.

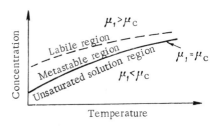

Fig. 6. Typical solubility diagram; μ_c is the chemical potential of a crystal; μ_1 is the chemical potential of the same crystal in solution.

The region of supersaturated solutions can be divided into two subregions: metastable (stable) and labile (unstable) solutions.

The width of the metastable subregion depends on the concentration of chemical, colloidal, and mechanical impurities. Carefully purified supersaturated solutions may be stored for years and can withstand supercooling of many tens of degrees [Akhumov (1960)]. The formation of nuclei is usually observed when the supercooling amounts to several degrees. Single crystals grow only in the metastable subregion near the solubility curve. (The reasons why supersaturated solutions are divided into metastable and labile states will be discussed later.)

A very important characteristic of the state of a crystallization medium is the degree of its deviation from equilibrium. This deviation is the driving force in crystallization, and is governed, as in other physicochemical processes, by the difference between the free energies of the system in the initial and final states. The process proceeds in the direction which results in a decrease of the free energy of the system. The free energy of a mole of a substance is known as the chemical potential μ.

For a dissolved substance in a solution in which the interaction between the particles of the solute is unimportant, we have

$$\mu = \mu_0 + RT \ln C,$$

where μ_0 is the chemical potential of the substance when its concentration in the solution is equal to unity (the absolute value of the chemical potential is unknown but this is usually unimportant); R is the universal gas constant, equal to 8.31 J·mole^{-1}·deg^{-1}; T is the absolute temperature, in °K; C is the concentration of the solute in the solution.

If a solution is saturated at a concentration C_0, i.e., if a solution is in equilibrium with a crystal, the chemical potential of the solute is equal to the chemical potential of the crystal. However, if some characteristic of the system (temperature, pressure, or concentration) is altered, the chemical potentials of the crystal and of the solute change as well. The specific relationship, for example, between the temperature and chemical potential of each component of the system, is responsible for the difference between the chemical potentials of a substance in the crystalline state and in solution. This difference is defined as follows:

$$\Delta \mu = (\mu_0 + RT \ln C) - (\mu_0 + RT \ln C_0) = RT \ln \frac{C}{C_0}.$$

The above difference is similar to the difference of levels of a liquid in two vessels or the difference of electrical potentials. When the particles of the dissolved substance interact with one another (i.e., if the solution is not ideal), the concentrations should be replaced with activities [Robinson and Stokes (1959)]. Unfortunately, the activity coefficients, representing the degree of departure of an actual solution from its ideal state, have been determined only for a very limited number of substances and mostly for unsaturated solutions. The growth of crystals of KNO_3, $NaBrO_3$, and other compounds, which are uniform when examined visually, is found to take place at values of $\Delta\mu$ equal to several units or a few tens of joules per mole.

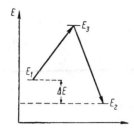

Fig. 7. Heat and energy of activation for a phase transition; ΔE is the heat of phase transition (heat of crystallization).

These values are very small compared with the values of $\Delta \mu$ in ordinary chemical reactions (thousands and tens of thousands of joules).

In practice, when it is not necessary to compare the rates and other features of crystal growth under different conditions, it is usual to employ simpler expressions to estimate the driving force of the crystallization process. This force is called the supersaturation. There are many ways of defining the supersaturation but their common feature is that this quantity is not expressed in terms of the energy. A quantity used very frequently is the relative supersaturation $\Delta C/C_0 = (C - C_0)/C_0$.

In practice, one often uses the supercooling of the solution. The supercooling is the difference between the saturation temperature of a given solution and the crystallization temperature ΔT.

The presence of a force of itself does not mean that a process will take place or that it will take place at a particular rate. The final result depends on the forces resisting the process. The particles in a liquid, in a crystal, or on its surface are "potential wells" and they are bound to the surrounding particles. In order to jump to the next equilibrium position (by diffusion in a solution, or from a solution to a crystal), a particle must break some of its bonds with its neighbors, i.e., it must overcome an energy barrier. This is shown in a schematic manner in Fig. 7. E_1 is the energy level in a supersaturated solution; E_2 is the energy level in a crystal. The height of the barrier, $E_3 - E_1$, is known as the heat of activation; E_3 is known as the activation energy. The heats of activations can be determined from the rates of growth at various temperatures [Zakhar'evskii (1963)].

The question arises how the process can take place at all if there is an energy barrier. The answer is that at any particular temperature, at any moment, the energies of different particles are different. The temperature simply represents the average value of the energy of all the particles. There are always local spontaneous deviations of the energy, concentration, and other macroscopic properties in a system from the average values; these deviations are limited to microscopic regions and they are known as fluctuations (for example, the well-known Brownian motion is the result of pressure fluctuations in a liquid or a gas). Due to these fluctuations, some particles temporarily have sufficient energy to overcome an activation barrier. The lower the average energy of the particles, the smaller is the number of particles which can overcome such a barrier per unit time and the slower the relevant process. When the temperature is increased, not only the average energy of the particles becomes larger, but also the magnitude and frequency of the fluctuations increase and, therefore, the number of jumps across a barrier becomes greater and all the molecular processes are accelerated.

The spontaneous appearance of tiny crystals (known as nuclei) in a strongly supercooled liquid is due to such fluctuations.

Near the surface a crystal has an excess of energy above the value of the energy in the interior and this excess is known as the surface energy. The greater the difference between the nature of the bonds in each of two phases in contact, the weaker are the bonds between the particles at the phase contact boundary (a complete absence of interaction at the boundary is impossible). Consequently, the greater the difference between the nature of the bonds in the two phases, the larger is the surface energy, and conversely. Therefore, to form a boundary separating a crystal from its ambient medium, we must perform work — the work of formation of a nucleus of a new phase. Even in unsaturated solutions, fluctuations of the concentration give rise to regions extending over several ions or molecules, in which the particle positions

TABLE 3

Compound	T_{mp}, °C	T_{bp}, °C
H_2Te	−48	−1.8
H_2Se	−64	−42
H_2S	−82.9	−61.8
H_2O	0	100

are the same as in the corresponding crystal phase. In a supersaturated metastable solution, the number of such regions and their dimensions are much larger. However, these regions are unstable and they disperse under the action of fluctuations of opposite sign. Such nuclei are called subcritical. If the supersaturation is increased still further so that the labile solution region is reached, nuclei of supercritical dimensions may be formed as a result of any fluctuations. Such nuclei do not disperse again and new particles are captured by them. Thus, supercritical nuclei grow in size. If the nucleation of crystals in metastable solutions does take place, it is due to the fact that the solutions are usually not very pure and contain dust particles on whose surfaces favorable conditions exist for an ordered assembly of particles. The energy of formation of supercritical nuclei is thus reduced by the presence of dust particles [Gorbachev and Shlykov (1955); Shylkov and Gorbachev (1955)]. Systematic experimental investigations of the width of the metastable region were carried out by Akhumov (1960). He found that the width of the metastable region decreased when the temperature was increased.

In the metastable region a pure solution contains an excess of the solute, which is essential for a transition to a new state, but the fluctuations are not sufficiently large to form a stable phase boundary, i.e., a supercritical nucleus. However, it is possible to grow single crystals in the metastable region.

We shall now consider some important properties of water, which is the most widely used solvent of ionic crystals and many organic substances. The atom of oxygen in the water molecule is bound to the two hydrogen atoms in such a way that the hydrogen nuclei (protons) are immersed completely in the electron cloud of the oxygen. The distance between an oxygen nucleus and a proton in ice is 0.99 Å, and in water vapor it is 0.9568 Å.

The H−O−H angle in water vapor is 105°03', whereas in ordinary hexagonal ice this angle is tetrahedral (109°30'). The protons are responsible for the presence of the two positive electric charges on the surface of the almost spherical water molecule. On the opposite side of the water molecule, there are orbits of two electrons, which are negative charges. The charges on the surface of the water molecule are at the vertices of a slightly distorted cubic tetrahedron. These charges are ±0.17 e, where e is the electron charge. Such a distribution of charges is responsible for the dipole moment of the water molecule, which is 1.84 or 1.87 D.*

This structure of the water molecules governs the nature of the bonds formed by it. Since these bonds are due to protons, they are known as the hydrogen bonds [Sokolov (1955)]. The influence of hydrogen bonds on the physical properties of substances may be very considerable.

Table 3 gives [Chemist's Handbook, Vol. 1 (1962)] the melting and boiling points of H_2O and its analogs, which do not have hydrogen bonds. In the case of the analogs of water, these quantities are governed by weaker bonds in the solid and liquid states.

---

*D is the debye, a unit of dipole moment, equal to 10^{-18} charge·cm in the cgs esu system.

Had the H_2O molecules been free of hydrogen bonds, then, judging by the tendency of the change in the boiling and melting points of analog compounds, the melting point of H_2O would be $T_{mp} \approx -100°C$ and its boiling point $T_{bp} \approx -80°C$. We can see that the difference between the actually observed and the theoretical melting and boiling points is very considerable, and is due to the presence of hydrogen bonds.

The directional nature of hydrogen bonds makes them similar to covalent bonds and, therefore, like covalent crystals, substances with hydrogen bonds frequently have loosely packed structures. The best example of this is the structure of ice.

Crystals of ordinary ice (known as modification I), form a hexagonal lattice which is stable, at 1 atm, down to $t = -70°C$. The structure of ice I is similar to that of tridymite (SiO_2) except that the proton oscillates between two atoms of oxygen, while in tridymite oxygen has a definite position between two atoms of silicon. The ice molecule is bound to its four nearest neighbors and this produces a loose structure with empty channels along the C axis. The cross sections of these channels at the widest points exceed the diameter of the water molecule. X-ray diffraction patterns indicate that when ice melts a new maximum appears corresponding to a distance of 3.47 Å from any molecule to the center of the nearest vacancy surrounded by six water molecules. This shows that the melting of ice is accompanied by the breaking of bonds of some molecules, which then occupy the channels in the ice structure. When the temperature is increased still further, the average coordination number increases to 4.9, which indicates a continuing destruction of the long-range order.

The fraction of molecules which lose their bonds with neighbors represents 9-15% at 0°C. The change in the density at the ice-water transition point is of the order of 10%. Since an increase in the density at the transition to the liquid state is due to a redistribution of molecules, we can assume that 10% of the molecules are shifted to the channels.

Several workers [cf. Danilov (1956)] have shown that the ability of a liquid to reach a supercooled state is a function of the difference between its structures in the liquid and solid states. The larger the differences between these structures, the greater is the tendency of a liquid to supercooling. In the case of water, it has been found that the degree of possible supercooling depends on the previous history, and after heating up to 30°C the supercooling is proportional to the previous heating. This is because the time for the establishment of the equilibrium state in water under these conditions is relatively long.

The nature of the motion of individual particles in a liquid is similar to that in solids, i.e., most of the time a given particle oscillates about a particular position. Samoilov (1957) has estimated the ratio of the number of vibrations of the water molecule about its equilibrium position to the number of jumps at 25°C. He found this ratio to be 1000 : 1. The path traveled by a particle in one jump is of the order of the interatomic distance and, therefore, there is no sustained inertial motion of a particle in a liquid.

The introduction of foreign ions into water (the process of solution) produces many effects whose magnitude depends considerably on the dimensions of these ions, their charge and concentration.

The process of solution is chemical, i.e., it involves a continuous interaction between the solute and solvent. The thermal, volume, and other effects are much weaker than those observed in ordinary chemical reactions because the strengths of the chemical bonds of various components of the solutions and the strengths of the bonds within each component are comparable. These effects are usually attributed to solvation, which is the term used to describe the formation of more or less stable complexes consisting of an ion (molecule) of the solute and a surrounding "screen" of solvent molecules.

One can distinguish the short-range and long-range solvation (in water, one speaks of hydration). The short-range solvation represents the permanent binding of solvent molecules to an ion. The long-range solvation represents changes affecting more distant solvent molecules (changes in the rate of diffusion, density, etc.).

It must be stressed that the interaction between an ion and a solvent is not limited simply to the nearest molecules. Lengyel (1959) has shown that the density of water increases at a distance up to 30-50 Å from the center of a foreign ion. The same conclusion has been reached by Mikhailov and Syrnikov (1960), who have found that in a solution all the water molecules are in the strong electrostatic field of foreign ions. Syrnikov (1958) is of the opinion that an anion interacts with water in the same way as water molecules with each other, while a cation interacts quite differently. Syrnikov concludes that the disturbance of the structure of water is mainly due to cations. In solutions from which solid hydrated crystals can be formed (such as lithium, calcium, aluminum, cerium, and thorium nitrates) the distribution of foreign ions tends to assume the form observed finally in hydrated crystals when the concentration of such ions is increased [Mathieu and Lounsbury (1949)]. The numerous investigations of Danilov et al. and of workers outside the USSR have established firmly that a uniform molecular mixing of components does not take place in alloys and solutions. Regions with the ordered structure of solids are retained in Pb−Bi, Sn−Bi, Sn−Pb, and Sn−Zn eutectics. This also applies to solutions of acetone in water [Danilov (1956)].

The quasicrystalline state of liquids is responsible for a number of effects which can be described as the polymorphism of liquids. Some of the properties of liquids, such as the surface tension, refractive index, density, etc., may change suddenly when external conditions are varied monotonically [Urazovskii (1956)]. These discontinuities are due to changes in the structure of liquids [Shakhparonov (1956), p. 485].

In the case of water, this sudden change is so large that, as reported by Mishchenko (1959), "we can, without exaggeration, say that water at 25-75°C and water near 0°C are two different solvents." Feates and Ives (1956) give somewhat higher values for the second melting point of water, namely, 30-40°C. Szent-Györgyi (1957) is of the opinion that a temperature of ~37°C is suitable for biological systems because at this temperature a transition takes place between the two forms of water and biological systems require these two different forms of water in many important life processes. Structural changes in solutions, observed when the temperature is varied, are associated with anomalies in the rate of growth of the faces of crystals of many substances from aqueous solutions, as discovered recently by Sipyagin and reported by Chernov (1965).

In supersaturated solutions, the structure (i.e., the properties similar to those of crystals) is even more pronounced than in unsaturated solutions and pure liquids. Thus, the viscosity of supersaturated solutions of $Ba(NO_3)_2$, Na_2SO_4, Na_2CO_3, etc. [Jha (1954)] decreases when the rate of flow is increased, while in other liquids the viscosity increases at higher flow rates. According to Jha, this is due to the relatively stable structure of supersaturated solutions and a perturbation of this structure by flow.

Thus, the ability of supersaturated solutions and other liquids to assume a crystal-like structure at temperatures not too far from their freezing points is a fairly common phenomenon.

We can see now that a liquid is, in many basic properties, more similar to a crystal than to a gas.

§4. Crystal Growth Mechanisms

The electric fields of particles on the surface of a crystal are not completely compensated. Consequently, those particles in a solution which are in the range of action of the electrical forces of the surface particles of a crystal are attracted by the surface and become attached to it. Thus, a crystal adsorbs particles from the surrounding medium. An adsorbed particle is either located permanently at a given point or, due to energy fluctuations, migrates along the surface and becomes permanently located at some other point or, having wandered on the surface for some time, may leave the surface (become desorbed) and return to the ambient medium [de Boer (1953)].

Surface diffusion during crystallization from solutions is of secondary importance, in contrast to a gaseous medium in which the arrival of particles at steps on the surface of a crystal (this will be discussed later) is due to the direct incidence of particles on the steps, as well as to the surface diffusion of adsorbed particles. If a crystal is in equilibrium with its ambient medium, the average numbers of adsorbed and desorbed particles (per unit time) are equal. However, if a particular substance is present in the ambient medium in a concentration higher than the equilibrium value, the number of adsorbed particles is greater than the number of desorbed particles and the crystal continues to grow. The rate of attachment of particles (the rate of crystallization) is, in general, proportional to the excess of the substance over the equilibrium value.

Because of the inhomogeneity of the surface of a crystal, we can distinguish the following cases of particle adsorption and migration (Fig. 8):

$$\text{a particle in a medium} \rightleftarrows \text{a smooth surface} + 0.030 \tag{1}$$

$$\text{a particle in a medium} \rightleftarrows \text{an edge of a crystal} + 0.056 \tag{2}$$

$$\text{a particle in a medium} \rightleftarrows \text{a step} + 0.113 \tag{3}$$

$$\text{a particle in a medium} \rightleftarrows \text{a kink in an incomplete plane} + 0.777 \tag{4}$$

$$\text{a particle on a smooth surface} \rightleftarrows \text{a step} + 0.083 \tag{5}$$

$$\text{a particle on a smooth surface} \rightleftarrows \text{a kink} + 0.747 \tag{6}$$

$$\text{a particle at a step} \rightleftarrows \text{a kink} + 0.664 \tag{7}$$

The numerical values given after each description represent the energy evolved in the process represented by the upper arrow (the forward process). The unit of energy is the work necessary to separate a pair of unlike ions in a gas. This calculation has been carried out by Kossel and Stranski [Honigmann (1961)] for a cubic F-type face of an ionic NaCl-type crystal and ions which are present in the host crystal, with the process taking place in a gaseous medium at 0°K.

Comparison of the relative adsorption energies (the first four transitions) shows that, on a smooth F-type face, the process most strongly favored by the energy considerations is the attachment of a particle in a kink of an incomplete plane. The next most favorable process is the attachment to

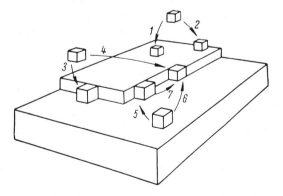

Fig. 8. Various cases of particle adsorption and migration on a growing crystal surface.

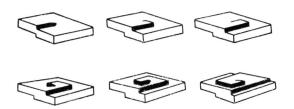

Fig. 9. Formation of a spiral step at the point of emergence of a screw dislocation.

a step (an edge of an incomplete plane). When an ion is adsorbed on a smooth surface, there is a considerable chance that it will be desorbed before it reaches a step and becomes attached to it.

The low value of the energy of attachment of a particle to a smooth face of the F-type means that the growth of a face of an ideal NaCl crystal cube stops when a layer (plane) is completed. In the case of a smooth surface and low supersaturations — as in the case of metastable solutions, in which crystallization does not start in the absence of a nucleus of sufficient size — the probability of fluctuation formation of a nucleus of a new layer (a two-dimensional nucleus) is very low. This is because of the high rate of motion of the front of a step compared with the rate of appearance of new steps and because the surface of a freely growing crystal has relatively flat faces.

NaCl sometimes forms octahedral faces of the K-type. Parallel octahedral faces represent alternate networks consisting of sodium or chlorine ions. Each adsorbed particles (which we shall assume to be positively charged) is located in a cavity between three particles in a lower layer, which is negatively charged. Thus, the adsorption of an ion is accompanied by the evolution of an amount of energy which is almost the same as in the case of adsorption in a kink of a step (incomplete plane) on the face of a cube. This results in a rapid and strong attachment of adsorbed ions and a rapid growth of an octahedral face. Therefore, this face frequently disappears altogether: it becomes completely overgrown. That is why NaCl-type crystals are cubes (in the absence of active impurities). Only at high supersaturations can we have a stable process of spontaneous formation of new layers on F-type faces. On the other hand, the growth of faces on which screw dislocations emerge can take place even at low supersaturations. This is due to the existence of a step which is always present and which begins at the point of emergence of a dislocation (Fig. 3). Thus, depending on the nature of the face, on the supersaturation, and on the chemical nature of the medium, the growth of faces may take place by the random attachment of new particles or by the growth of complete layers. Moreover, faces may grow by the attachment of three-dimensional nuclei. The mechanism of growth involving single particles is called normal [Cahn (1960); Cahn, Hillig, and Sears (1964); Borisov, Dukhin and Matveev (1964)]. The layer mechanisms are divided into growth by the attachment of two-dimensional nuclei and the dislocation growth mechanism. These two mechanisms are discussed in detail by Hirth and Pound (1963). The growth mechanism involving three-dimensional nuclei has been considered by Sheftal' (1958).

Dislocation Growth Mechanism

If we assume that a step due to a screw dislocation is initially straight (Fig. 9) and the rate of attachment of particles to the whole step is constant (the step grows at a constant linear velocity), we find that the angular velocity of various points in the step is not the same and the step twists into a spiral. This produces a cone-like projection (known as a growth cone) above the point of emergence of a dislocation (Fig. 10). In the case of slow dissolution, we can observe the reverse motion of a spiral step ending with the formation of a pit at the point of emergence of a dislocation [Kozlovskii (1958)].

At low supersaturations, the symmetry of the contours of the steps is the same as the symmetry of the faces [Frank (1952); Dukova (1960)]. The heights of the steps are not great and the growth cone, whose vertex angle is close to 180°, is practically invisible. The lateral surfaces of the growth cones sometimes form macroscopically smooth surfaces which reflect light.

Fig. 10. Spirals on the surface of a crystal: a) a polygonized growth spiral on the face of a $CoFe_2O_4$ crystal, ×500; b) a double spiral on $MnFe_2O_4$, ×500 (photomicrographs supplied by Sobolev).

When the supersaturation is increased, the distance between the steps decreases and the steepness of the cones becomes greater. The relief of the surface shows more contrast. The contours of the steps are then usually circular.

The contrast in the relief consisting of elementary ("repeatable") steps of the same height may be increased by the appearance of "kinematic waves of the step density" [Cabrera, and Vermilyea (1958); Chernov (1961)]. The appearance of such waves, which represent a higher density of these elementary steps, is due to characteristic features of the diffusion field near growth centers and the ends of such kinematic waves.

Although Fig. 11a does not show the elementary steps forming ring-shaped kinematic waves, the waves themselves are clearly visible and the growth center (vertex of the growth

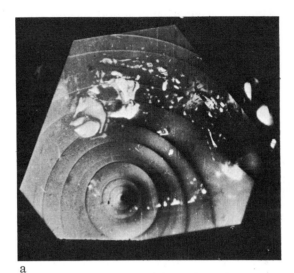

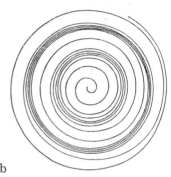

Fig. 11. a) Kinematic waves on the face of a sodium bromate tetrahedral crystal, ×35; b) schematic representation of the formation of concentric kinematic waves from elementary spiral steps.

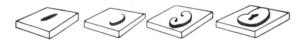

Fig. 12. Development of a pair of screw dislocations with the Burgers vectors of the same magnitude but opposite sign.

cone) can be found easily. In some cases, the kinematic waves can be observed even with the naked eye. The elementary steps can be observed using special fine methods for investigation of surfaces (they include electron microscopy and multibeam interferometry) [Verma (1953); Tolansky (1960)].

In some cases, the elementary steps themselves acquire a ring-shaped form. This is due to the effect of a pair of screw dislocations with the Burgers vectors of the same magnitude but opposite sign (Fig. 12).

Although the density of dislocations in ionic crystals ranges from 10 to 10^5 cm^{-2}, under stable conditions only a few centers (sometimes only one) are active on a given face. However, if we change the supersaturation, we find that the surface becomes covered with a larger number of small growth cones. After some time, we again end with a few growth centers but these centers are new and not those active before. When we return to the previous saturation level (after observing the multiple appearance of the growth centers), we find that the old centers become reactivated [Kasatkin (1964)]. These observations confirm the correctness of the theoretical conclusions [Frank (1950); Burton, Cabrera, and Frank (1951)] that the degree of activity of a growth-cone center (a group of dislocations) depends on the supersaturation.

In many cases at high supersaturations, the first dislocation (or the first group of dislocations) which appears along some direction is responsible for the rapid growth of a crystal along this direction before the appearance of dislocations along other directions. This is why, in some cases, we observe the growth of whiskers.

There are quite a few ways in which dislocations may appear in a growing crystal. Lemmlein and Dukova (1956) have observed the appearance of dislocations when the branches of a skeletal form intergrow (such skeletal forms will be discussed in §6) due to a slight misorientation of the branches during their free growth. Moreover, if a growing layer bypasses a dust particle located on the surface of a crystal, the distribution of crystal particles near this point is somewhat distorted. Therefore, when the edges of the layer close up around the dust particle a dislocation is formed. Figure 13 shows the formation of a dislocation with edge and screw components at its two ends emerging on the surface. In exactly the same manner, a dis-

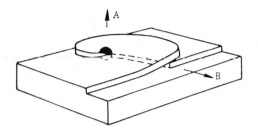

Fig. 13. Formation of a dislocation around a foreign particle on the surface of a crystal: A) direction on the axis of a screw dislocation; B) direction of the axis of an edge dislocation.

location is formed when layers join above cavities filled with liquid. Therefore, if a given growth pyramid contains liquid inclusions which are formed during growth (they are known as primary inclusions), dislocations emerge on the surface of the corresponding face above the inclusion. Secondary inclusions, which appear after the formation of a crystal, need not be accompanied by dislocations.

Growth and Dissolution Involving Two-Dimensional Nuclei

We have already mentioned that the probability of the formation of stable nuclei on a smooth and perfect crystal surface is very low. This probability is low right up to a relative supersaturation amounting to several tens of percent. The fact that the dislocation growth is often the dominant growth mechanism provides evidence of the low probability of the formation of stable two-dimensional nuclei in a wide range of supersaturations. During dissolution (again in a wide range of unsaturated concentrations) a perfect face does not dissolve, as demonstrated by Sears (1957). In Sears' experiments the faces of paratoluidine crystals subjected to a jet of unsaturated vapor did not evaporate (far from the edges) even when the jet was far from saturation (50%).

The theoretical dependence of the rate of growth on the supersaturation shown in Fig. 14 has been confirmed experimentally in the growth of crystals from a vapor. When a crystal grows from a solution, similar dependences are also observed but they do not necessarily indicate that the growth is due to two-dimensional nuclei. This is because a solvent may be adsorbed quite strongly to the faces producing a "screen," as in the case of colloidal particles, and although the growth actually takes place by the dislocation mechanism, the form of the dependence suggests misleadingly the growth involving two-dimensional nuclei. Examination of the relief of a face and its morphology are also insufficient to determine whether the growth is due to two-dimensional nuclei, because the thickness of these nuclei is equal to the lattice period. Therefore, in order to determine the growth mechanism it is necessary to use optical systems with a vertical resolving power of a few angstroms. The formation of layers at corners and edges is usually regarded as a sign of the growth involving two-dimensional nuclei. However, the formation of layers at corners and edges is frequently due to the special properties of the diffusion field near a crystal in the case of the dislocation growth mechanism [Dekeyser and Amelinckx (1955), p. 136]. Therefore, the growth at edges and corners cannot be regarded as a definite indication of the two-dimensional nucleation. A probable sign of the two-dimensional nucleation is the scale of oscillations of the rate of growth of a face with time under steady-state conditions. It has been shown [Kasatkin (1965)] that in the case of sodium bromate such oscillations reach their maximum amplitude at a certain value of the supersaturation and then decrease. At low supersaturations, sodium bromate crystals grow by the dislocation mechanism. When the supersaturation is increased, the formation of thick layers at the edges and corners of a crystal is observed and the appearance of these layers coincides with a decrease in the oscillations in the rates of growth. The rate of growth becomes more stable because the dislocation mechanism becomes secondary and two-dimensional nuclei are formed at a relatively uniform rate.

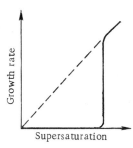

Fig. 14. Theoretical dependence of the rate of growth of a perfect crystal on the supersaturation in the case of two-dimensional nucleation

The curve in Fig. 14 shows the limit reached in an ideal case. In real systems, two-dimensional nuclei begin to make a contribution even at low supersaturations because of the presence of impurities on the surface, which act as nuclei of a new layer. By analogy with the three-dimensional nucleation, the range of supersaturations, in which the formation of two-dimensional nuclei has a low probability, should become narrower as the temperature rises.

It is known that the anisometry of a crystal increases when the supersaturation is increased and this can be explained by two-dimensional nucleation. Since different faces have different potential reliefs, the supersaturation at which the two-dimensional nucleation begins is different for different faces. At a given supersaturation, the rate of formation of two-dimensional nuclei on different faces may be so different that a crystal may develop a strong anisometry.

The most convincing demonstration of the two-dimensional nucleation in crystallization of dislocation-free silver crystals in an electric field was described by Kaishev (1966). Crystals were grown in capillaries in which it was quite easy to obtain a dislocation-free surface because the dislocations emerged on the lateral surfaces of a crystal. When a suitable voltage was applied to a system consisting of a crystal and a solution of silver nitrate and no dislocations had emerged on the end of the crystal, it was possible to observe discontinuous passage of an electric current in the form of separate pulses. It had been established that the current passed only at the moment of crystallization and the current should have been continuous in the case of dislocation growth. By measuring the amount of charge flowing in one pulse, it was found that the charge represented the number of electrons necessary for the reduction of the quantity of silver which would cover the end area of the capillary-grown crystal with a monatomic layer.

Growth by Attachment of Three-Dimensional Nuclei

Frenkel (1945) demonstrated that in a saturated vapor there are considerable numbers of complexes containing more than 1000 molecules each. When the supersaturation is increased, the amplitude of the fluctuations in a solution increases and its microheterogeneity becomes greater. There are grounds for assuming that a strongly supersaturated solution is similar to colloidal solutions. Therefore, at high supersaturations in the metastable region, a crystal may grow not only due to two-dimensional nucleation on the face but also due to the attachment of three-dimensional particle aggregates present in the solution (they are known as embryos, prenuclei, submicron particles, or complexes). When the supersaturation is increased, the importance of this growth mechanism should increase. At high supersaturations, near the labile region, this mechanism may become dominant. Gol'dshtein (1941) described the growth of sodium bicarbonate crystals of up to 100-150 μ in size by the aggregation and the subsequent joining of hundreds of crystallites. Peibst and Noack (1962) showed that the rate of formation of potassium chloride nuclei and the rate of growth of KCl crystals from aqueous solutions obeyed the same law at high supersaturations. Hence, they concluded that the growth of KCl crystals was due to the joining together of these three-dimensional aggregates.

Because a face is usually imperfect and has strongly adsorbed particles, the orientation of captured embryos is not always perfectly parallel [Shubnikov (1935)]. The activation energy for the reorientation of the embryos is relatively large because of their large mass and, therefore, a crystal which grows by such a mechanism may be badly formed and may have a mosaic or block structure. It is well known that the higher the supersaturation, the more pronounced is the block substructure of a crystal. Just as between ideal metastable supersaturated solutions and colloidal solutions there are many intermediate steps, so there are many intermediate forms of crystals between crystals of good quality free from blocks and colloidal completely disordered precipitates.

A very important factor which governs the degree of ordering at the crystal-ambient medium boundary is the rate of the growth. The higher the rate of growth, the higher is the degree of disorder (other conditions being equal). Moreover, the more viscous the solution, the easier is the formation of crystals with block substructure. This is because the diffusion of large particles in viscous media is retarded. For this reason, the so-called spherulites (crystalline intergrowths with radial structure) (Fig. 48) are formed mainly at high supersaturations and in viscous media. The formation of spherulites is closely related to the formation of dendrites, which are tree-like crystals with disoriented branches. There are intermediate forms between dendrites and spherulites [Keith and Padden (1963)].

We must mention that in any growth mechanism at comparable crystallization rates, less perfect crystals are obtained for more complex molecules. For this reason, spherulites and dendrites are formed in crystallization of polymers [Mandelkern (1964)].

§5. Role of Strongly Adsorbed Impurities

Let us assume that at a given moment practically all types of adsorbed particle are present on the surface of a crystal.

During the growth of the crystal there is a continuous competition for space on the surface between the particles forming this crystal and the impurities, which may be molecules of the solvent or of some other substance.

de Boer (1959) cites a formula, which unfortunately has not been checked experimentally, for the case of adsorption from a vapor:

$$\frac{Q_1}{Q_2} = \frac{P_1}{P_2} \exp \frac{U_1 - U_2}{kT}.$$

Here, Q_1 and Q_2 are the areas occupied by particles of types 1 and 2; P_1 and P_2 are the pressures of these particles in the ambient medium; T is the absolute temperature in °K; k is the Boltzmann constant; U_1 and U_2 are the energies of adsorption of these particles on a given surface (in the case of crystallization from a solution, the vapor pressures are replaced with the concentrations of the substances in the solution).

The correctness of the qualitative conclusions which follow from this expression indicate that, at least in the first approximation, the expression is correct. It follows from it that the ratio of the areas occupied by impurities is governed primarily by the energies of adsorption and only secondarily by their concentrations in the ambient medium. The electrical fields of particles constituting the various crystallographic faces are compensated to different degrees and the resultant fields are different. Therefore, the energy of adsorption of a given particle on different faces is different; in other words, the adsorption on different crystallographic faces is specific to those faces.*

The adsorption energies, governed to a considerable degree by the nature of the chemical binding between adsorbed particles and the surface, can vary considerably. When the temperature is increased, the strength of binding decreases. However, even at high temperatures, the strength of bonds may be so high that, in some cases, an adsorbed particle cannot be detached from the surface without damaging the structure of the substrate. For example, it has been found that the adsorbed water cannot be removed completely from the surfaces of NaCl, $BaCl_2$, and CaF_2 by heating in vacuum. During such at about 500°C, chemical reactions take place in which the corresponding acids are evolved [de Boer (1959)].

The average residence time of an adsorbed particle on the surface from the moment of adsorption to the moment of desorption is given by

$$\tau = \tau_0\, e^{\frac{U_0}{kT}},$$

where τ_0 is the period of complete oscillation of the particle (10^{-13} sec) and U_0 is the average adsorption energy of a given particle on a given face.

If the heat of activation for the jump of a particle from one position to another is high, then, in practice, an adsorbed particle will not migrate (diffuse) along the surface. This is likely to be true on the surface of a crystal in a solution. In this case, τ is the time during which a site occupied by a foreign particle cannot be filled by a particle forming the host crystal. Moreover, because of the distortion of the potential field of the surface, normal filling of the nearest sites is not possible during the residence time of an impurity. Two situations may arise. If a foreign particle is adsorbed very strongly (i.e., the value of τ is large), then foreign particles in front of a moving step can be regarded as a sort of fence, which is known as a Cabrera fence. For a step to move without difficulty, the distance between impurity particles should be larger than a certain value. This value depends on the temperature and supersaturation, and can be comparable with the size of a two-dimensional critical nucleus. Impurities in the form of large organic molecules with large values of U_0 behave in this way. However, if the adsorption energy is small, the kinks in the steps play the dominant role. The adsorption of impurities at the kinks usually reduces the rate of growth. If the end of a step is parallel to a smooth face (in this case, the number of kinks is small), a very small number of strongly adsorbed impurities can block the growth of a step and, therefore, reduce the rate of growth of a face, and sometimes stop the growth altogether. Usually, strongly adsorbed impurities act differently on the rates of growth of different faces, i.e., they alter the habit of a crystal.

An elementary analysis of de Boer's expression allows us to estimate the influence of temperature on the change in the ratio of the rates of growth of faces due to pure adsorption effects. Assuming that U_1 and U_2 represent the adsorption energies of a given impurity on different faces, we can estimate roughly the ratio of the areas occupied by this impurity on dif-

---

* Physicochemical investigations usually yield only one value of the adsorption energy for a given pair of substances. This energy is the average value of the adsorption energies of different faces and of different points on these faces; it includes also the interaction of all the adsorbed particles with each other.

ferent faces. Since the concentration of an impurity on different faces is, on the average, the same, we find that

$$P_1 = P_2 \text{ and } \ln\frac{Q_1}{Q_2} = \frac{U_1 - U_2}{kT}.$$

It follows from the above expression that the influence of temperature on $\ln(Q_1/Q_2)$ should be most noticeable at: 1) low temperatures, and 2) high values of the difference between U_1 and U_2, i.e., when the adsorption of an impurity on different faces is highly specific.

Numerous examples of the strong influence of negligible amounts of impurities (one or fewer particles per million) on the growth of crystals have been reported, particularly by Buckley (1951). The reported strong change of the habit provides a very rough method for investigating the influence of impurities on the growth of a crystal. The point is that an impurity may either reduce the rate of growth of all the faces by approximately the same amount (nonspecific effect) or reduce the rates of growth of some of the faces (specific effect). Only the latter results in a change of the habit of a crystal. In the case of a weak specificity, the influence of an impurity may not be observed at all, while the absolute rates of growth may be strongly affected.

In the presence of impurity islands on the surface, the contours of the steps become twisted or sometimes sawtooth-shaped. In such a case many microscopic inclusions appear in a crystal and, in extreme cases, opaque crystals of milky appearance are obtained. For example, lead nitrate may grow from an aqueous solution in the form of opaque porcelain-like crystals because of the etching of the surface by the less soluble basic salt $Pb(OH)NO_3$.

The simplest and most reliable method for estimating the adsorption capability of the components of a solution is the Paneth's rule [Lemmlein (1948)] which predicts that a crystal adsorbs most strongly those ions which produce the least soluble compound when combined with ions in the host crystal. Thus, we can predict that a K^+ ion in a solution of $NaBrO_3$ will be strongly adsorbed on $NaBrO_3$ because of the low solubility of $KBrO_3$. Consequently, this ion should affect strongly the growth of sodium bromate, while the presence of the same amount of Cl^- and Br^- ions should have hardly any influence on the growth process.

There are considerable differences between the solubilities of acid and normal salts and of normal and basic salts. Therefore, a change in the value of the pH of a medium shifts the chemical equilibrium in a solution in accordance with the same Paneth's rule and this affects the adsorption process. In the case of the formation of less soluble compounds, a change in the pH reduces the rate of growth of a crystal. Thus, potash alum grows more slowly when OH^- ions are introduced into the solution [Portnov and Belyustin (1965)]. This is expected, since the introduction of this group should produce poorly soluble aluminum hydroxide. Correspondingly, if the growth is decelerated by the formation of a poorly soluble compound on the surface and the change in the acidity shifts the reaction in the direction of the formation of a more soluble compound, the rate of growth of a crystal should increase. However, even if the introduction of H^+ and OH^- does not alter the adsorption mechanism, a change in the acidity may still affect the rate of growth of a crystal for a different reason, which will be discussed later.

Buckley (1951) reports that organic dyes are strongly adsorbed. The cause of this lies in the chemical unsaturation of these compounds, i.e., the presence of an odd number of electrons in their molecules [Kondrat'ev (1959), p. 269], which makes easy the adsorption of these molecules.

A large number of organic compounds (for example, amines) are known, which, because of the presence of polar groups, are adsorbed easily on the surfaces of various inorganic substances. They are known as surface-active substances or surfactants [Encyclopedic Physics Dictionary, Vol. IV, p. 56]. These substances must be borne in mind in the search for impurities affecting the growth of crystals.

In some cases, we find that if a compound is insoluble and decomposes, even at a low rate, then crystals grow slowly and have many defects. This is because of the high activity of the reaction products at the moment of reaction, i.e., because such a compound is present in the form of radicals in a solution.

If the particles of some impurity have the same charge as the particles of the host crystal and differ little from the latter in their size (an isomorphous impurity), such an impurity is adsorbed by a growing crystal to practically the same extent as the components of the host crystal. Isomorphous impurities have little influence on the rate of growth of faces, i.e., on the habit of crystals.

Some impurities present in a solution may form their own crystals on a host crystal, and these impurity crystals may have a regular configuration. This phenomenon of oriented growth on a substrate is known as epitaxy and the impurity is called epitaxial. Many experimental investigations have demonstrated that the cause of epitaxy is the similarity of the distribution of particles and the distances between them at the contact surface between the substrate and the epitaxial layer. This rule is not universal and the causes of epitaxy are somewhat more complex, so that at present we do not have a complete system of criteria for the prediction of epitaxy [Chistyakov and Lainer (1966); Palatnik and Papirov (1964)]. It is very important to point out that the growth of a crystal may be affected even when an impurity is not precipitated in the form of crystals. The addition of negligible amounts of potassium ferrocyanide to a KCl solution affects considerably the kinetics of crystallization and the quality of the KCl crystals. It has been established that this impurity is epitaxial [Neels and Steinicke (1963)]. Because of the strong specificity of the epitaxy, the impurities adsorbed by this mechanism affect in different degrees the rate of growth of different faces. Therefore, the effect of such impurities on the habit of a crystal is usually strong. Geometrical (epitaxial) relationships between the substrate and adsorbed substances are important at the molecular level and they play an important role in the well-developed multiplet theory of catalysis [Balandin (1963-1964)]. In general, many relationships which apply to catalysis should have their counterparts in crystal growth. The processes of catalysis and crystal growth are very closely related.

The high activity of a crystal, in relation to a layer of similar structure in contact with it, may give rise to crystal phases which are not stable under given thermodynamic conditions in the absence of a substrate. This is demonstrated by the formation of wüstite (FeO) on the surface of an iron crystal at room temperature although the stability range of FeO lies above 600°C [Dankov and Shishakov (1949)]. The same phenomenon is observed in the dependence of the temperature of the transition of ice from one modification to another on the properties of the substrate. The difference between the transition temperature with a substrate and those without a substrate may reach 25 deg C [Lisgarten and Blackman (1956)].

Deryagin, Karasev, and Zorin (1954) are of the opinion that the orientation is firmly established in a layer 10^{-5}-10^{-6} cm thick. Gol'danskii and Chirkov (1947) determined the maximum thickness of adsorbed films of ethyl acetate, alcohol, water, and acetic acid on mica, and found that this thickness was $(1-3) \cdot 10^{-6}$ cm. For a 20% solution of KCl, this thickness is considerably less: $2 \cdot 10^{-7}$ cm.

Henniker and McBain [Szent-Györgyi (1957)] and Akhmatov (1963) have concluded that near a solid-liquid boundary there is a liquid layer of strongly modified orientation, having a thickness of the order of tens and hundreds of molecules.

Abdrakhmanova and Deryagin's work (1958) on the surface activity of quartz in the presence of adsorbed layers is particularly interesting. They measured the conductivity of films of eight substances, including water, adsorbed on the surface of a carefully cleaned quartz crystal. They found that the admission of vapors of the investigated substances into an evacuated chamber containing a quartz crystal increased the surface conductance of a crystal considerably. After 1.5-2 min, the value of conductance reached its maximum and then, during the next 5-10 min, it decreased fairly rapidly. Beyond this stage, the conductance remained constant. Discussing the causes of this effect, Abdrakhmanova and Deryagin write: "Obviously, initially a nonequilibrium adsorbed layer is formed on the surface of quartz and the molecules in this layer do not have an equilibrium orientation and they do not necessarily occupy sites with a maximum adsorption energy (because of inhomogeneity of the surface); this is responsible for the higher conductance. The next stage is the orientation of adsorbed molecules and their migration to sites with higher adsorption energies. The process of orientation spreads from lower to upper layers so that an adsorbed film is formed with strongly attached molecules and an ordered structure ... ."

A solvent may be adsorbed strongly on the faces of a growing crystal. This is indicated by very large differences between the habit of organic crystal substances grown from different solvents [Wells (1946)] and the results of studies of the growth of crystals from aqueous solutions [Lewin (1955); Kleber (1957); Petrov (1964)]. A solvation shell (screen), whose existence has been established for single ions, molecules, and colloidal particles, may exist also on the surface of a crystal. It has been found that the energy of adsorption of water from the gaseous phase on the surface of potassium chloride at 0°C is 50 kJ/mole (this value is considerably higher than the usual hydrogen bond energies − cf. §1). When epitaxial relationships apply between the face of the substrate and the face of a solvent crystal, the solvation shells are stronger since the ordered nature of the first adsorbed layer of molecules favors the spreading of the ordered state into the liquid. Thus, it has been shown [Palmer, Cunliffe, and Hough (1952)] that the permittivity ε of water films 2-3 μ thick, compressed between mica plates, is 10 at room temperature, while ordinarily ε of water is ≈ 80. The frequency dependence of ε of water under these conditions has been found to be similar to its dependence for ice, indicating a high degree of order in such a water film. In general, mica is epitaxial in relation to ice. It would be very interesting to determine ε of water in contact with other mica crystals having different cleavage-plane parameters, in order to determine the relationship between the degree of structural similarity between mica and ice, and the degree of ordering of water. The consequences of the formation of thick polymolecular layers of an impurity on the surface of a growing crystal are important also in the problems discussed in the next section. These effects are of cardinal importance for the understanding of the active role of the crystal surface, which alters the properties of a medium at the crystal-solution boundary.

Impurities are unavoidably captured by a crystal (absorbed in the crystal). Apart from the cited factors, which affect the adsorption, the penetration of impurities into a crystal is governed by the rate of growth and the rate itself depends on the adsorption of the impurities. Unfortunately, the experimental data on the dependence of the penetration of impurities on the rate of growth are scant. The results which are available are simply isolated examples. In obtaining these results, it has been usual to ignore the fact that an impurity penetrates different faces in different amounts, that impurities are captured by a crystal together with solution inclusions; moreover, the experimental conditions have not been kept steady, etc. However, it is known that isomorphous impurities are captured by different faces in different amounts.

The amount of an impurity captured by a crystal is usually represented by the segregation coefficient, which is defined by the following formula for growth from a solution:

$$k = \frac{Q_{mS}}{Q_{MS}} : \frac{Q_{mL}}{Q_{ML}},$$

where Q_{mS} is the amount of an impurity in a crystal; Q_{MS} is the amount of the host substance in a crystal; Q_{mL} is the amount of the impurity in a solution; Q_{ML} is the amount of the dissolved host substance.

For weakly adsorbed impurities, k is less than unity but for impurities which are strongly adsorbed and captured by a crystal, k can exceed unity.

The distribution of a captured impurity in a crystal is nonuniform. Because of the different adsorption energies, an impurity penetrates in different amounts the growth pyramids of the different crystallographic faces [Lemmlein (1948)]. A growth pyramid is a part of a crystal which is formed by the growth of a particular face. The base of this pyramid is the face of the crystal and its apex is the original growth point, while the walls are boundaries with neighboring growth pyramids. Because of the nonuniform capture of an impurity by different growth pyramids, the distribution of an impurity in a crystal is nonuniform. For this reason, the physical properties of the pyramids may differ very considerably. Thus, if we cut a crystal normally to its faces, subsequent polishing and etching reveals that different pyramids are dissolved at different rates. Moreover, such pyramids differ in their hardness and refractive index, sometimes in their color, etc. The nonuniform distribution of an impurity in growth pyramids results in a sectorial structure of a crystal.

Moreover, within the same sector, an impurity captured in different amounts at different stages of growth of a crystal gives rise to a zonal structure. The principal cause of the zonal structure is the dependence of the segregation coefficient on the kinetics of the crystallization processes. Changes in the crystallization conditions and spontaneous fluctuations in the rate of growth of faces result in changes in the composition of a crystal during its growth. Finally, an impurity is captured nonuniformly across the surface of a given face, depending, for example, on the distribution of steps and kinks on the surface [Lemmlein (1948)].

All these differences between the concentration of impurities in different parts of a crystal produce unit cells with different parameters. This phenomenon is known as "heterometry" [Shternberg (1962)] and occurs widely in natural and synthetic crystals. The heterometry produces local stresses in a crystal which may be manifested in various ways. Thus, Tiller (1958) has shown that when an impurity is captured nonuniformly by a particular growth pyramid, dislocations may be generated. Treivus, Petrov, and Kamentsev (1965) have demonstrated that dislocations can also be generated when the stresses are distributed between neighboring growth pyramids. Obviously, impurities are the main cause of the formation of dislocations during the growth of a crystal from a solution or a vapor, if the medium does not contain an appreciable amount of dust and if inclusions are not formed. If a material is not plastic, such stresses are relieved by the bending of a crystal or by cracks (which are usually oriented in a regular manner); block substructure may appear as well. In some cases, these cracks affect the growth pyramid of only one crystallographic form. This has been observed in the growth of Rochelle salt containing tin and other impurities [Shternberg (1962)]. The rate of growth of a crystal depends on the number of defects in it. Crystals containing large macroscopic defects (particularly cracks) grow much faster than crystals which are uniform to the naked eye. Thus, we frequently observe the following sequence: adsorption of impurities, penetration of impurities into a crystal, formation of defects, and increase in the rate of growth [Sheftal' (1958)]. Naturally, an isomorphous impurity which may be captured by a crystal in any amount does not produce defects. Any other impurity disturbs the structure of a crystal and the greater the difference between the impurity and the host crystal, the greater is the disturbance. In some (rare) cases, the rate of growth may also increase when adsorbed particles favor the formation of two-dimensional nuclei.

Adsorption is only one of the processes which governs the growth of crystals. Another important factor is diffusion. Diffusion is responsible for the transport of matter to a crystal and for the dissipation of the heat of crystallization. In solutions, the transport of matter is more important than the dispersal of heat. In the crystallization from pure melts, only heat transport is observed but no mass transfer.

§6. Volume Diffusion During Growth of Crystals

Diffusion is the distribution of the concentration of a substance by a random motion of individual particles; it is due to the presence of a gradient of the chemical potential in the system.* Diffusion always reduces this gradient. It is frequently assumed that the driving force in diffusion is a concentration gradient. This is incorrect. For example, the concentration in a solution at various points may be the same but if a temperature difference is established, a gradient of the chemical potential is produced and diffusion takes place, resulting in a redistribution of the concentration. This process is known as thermal diffusion or the Soret effect. The diffusion continues until the chemical potentials become equal.

In the absence of an appreciable temperature difference, we can ignore the chemical potentials and consider diffusion simply as a function of the concentration in a system. Then, the amount (dM) of matter transported in molecular diffusion during a time interval (dt) is, according to Fick's law:

$$\frac{dM}{dt} = D \frac{dC}{dx} dS,$$

where D is the diffusion coefficient; dC/dx is the gradient of the concentration C; dS is the area through which transport takes place.

We can distinguish three different forms of diffusion during the growth of crystals from solutions or during the dissolution, etc. of crystals [Zdanovskii (1956)]; they are discussed below.

1. <u>Molecular Diffusion (Crystallization in a Solution at Rest)</u>. Such diffusion is observed in viscous media and at low supersaturations, as well as in the growth of crystals in thin films of liquids or in capillaries.

A layer of solution near a crystal within which the concentration of a solution changes and diffusion takes place is known as the boundary diffusion layer.

In molecular diffusion the transport of matter to a crystal is slower than under other diffusion conditions. The thickness of the boundary layer increases with time and the concentration gradient gradually decreases. Therefore, the rate of growth decreases with time. Matter reaches projecting parts of a crystal (corners and edges) in amounts larger than those in the centers of various faces; therefore, supersaturation gradients appear along faces. If these gradients and the dimensions of the crystal are small, a flat-faced crystal is obtained. When the supersaturation gradients and the dimensions of the crystal are increased, we reach a position when the supply of matter to the middle parts of the faces decreases so much that the layers growing from opposite edges do not meet. A cavity appears on the face. Subsequent layers cover this cavity (because of the partial re-establishment of the supersaturation at the center of the face) and an inclusion of the solution is captured by the crystal. This process may be repeated many times. In this way, a series of inclusions one above the other may be produced (Fig. 15). At still higher gradients or when the dimensions of the faces are large, the conditions for the closing over of such cavities may not be reached. Thus, crater-like

---

*A gradient is defined as an increment of a function (in this case, dμ) in an infinitely short distance dx, along the direction of the most rapid variation of the function.

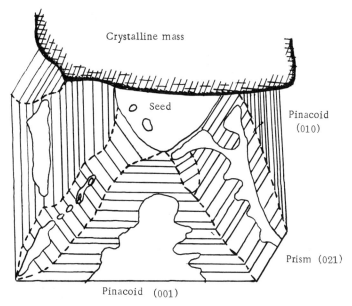

Fig. 15. Growth of a potassium nitrate crystal recorded with a cine camera. Crystal size 1.5 mm. Superimposed contours of an inclusion in a growth pyramid on the (021) face represent changes in the shape and size of this inclusion during growth. It is interesting to note that the opposite faces of a (010) pinacoid grow at different rates. Moreover, although the external conditions (temperature and supersaturation) are constant, the faces are seen to grow nonuniformly.

Fig. 16. Corner skeletal form of ammonium chloride. ×20.

hollows are formed on the faces. A crystal which retains its edges and parts of its faces near the edges is known as an edge skeletal form. At still higher gradients or dimensions of the faces, the difference between the supply of matter to the corners and edges becomes important and the edges break up. The deposition of matter mainly at the corners results in the formation of branched crystals which can be called corner skeletal forms (Fig. 16). Crystallographic orientation of all branches of a corner skeletal form is the same. Unlike dendrites, skeletal forms are single crystals.

In Fig. 15, the right-hand face (021) is approximately twice as wide as the left-hand face ($0\bar{2}1$). A crater-like hollow has been observed on the right-hand face throughout its growth and the face has not closed over this hollow. The left-hand face shows a series of inclusions. Thus, Fig. 15 illustrates the relationship between the dimensions of a face and its growth characteristics.

The lower the rate of diffusion, the greater is the difference between the conditions of supply of matter to different parts of the same face. Therefore, under molecular diffusion conditions, uniform crystals can be grown at lower supersaturations than under other conditions.

2. Natural Convection. Under natural convection conditions, a solution overcomes its viscosity and moves relative to a crystal. Natural convection is due to the difference of densities of a solution near a crystal and far from it. The difference between the densities is due primarily to a change in the concentration of a solution during growth or the dissolution of a crystal; and, secondly, due to the absorption or evolution of heat in these processes. During the growth of a crystal, matter rises because the solution near a crystal is less dense as a result of the reduction in the concentration, and the temperature is higher because of the evolution of the heat of crystallization. When a crystal is dissolved, the direction of the motion is opposite (downward). Under these conditions, the molecular diffusion (which is always present) is supplemented by the more energetic convective transport of matter.

The time interval during which a boundary diffusion layer is formed represents the nonsteady-state conditions. During this initial period, the rate of growth varies considerably. After this period, if the supersaturation in a solution is kept constant, steady-state conditions are obtained. The time needed to establish a boundary layer or a natural convection layer is between a few minutes and tens of minutes for crystals measuring 1-5 mm. The thickness of the boundary layer depends on the difference between the densities of different parts of the solution (i.e., on the rate of growth of a crystal), the viscosity of the solution, and the dimensions of the crystal. The presence of some walls near the crystal and the orientation of the crystal itself affect the nature of the convection currents and the thickness of the boundary layer at different faces. The greater the departure of the orientation of a face from the vertical position, the weaker is the convective current and the weaker the supply of matter, particularly in the case of faces turned upward. Under natural or free convection conditions, the thickness of the boundary layer formed near the crystal is of the order of tenths of a millimeter.

Because the boundary layer is thinner than in molecular diffusion, the concentration gradients (normal to the surface of a crystal) are higher. This results in an increase in the rate of transport of matter to the surface of the crystal and correspondingly increases the rate of its growth under these conditions.

The supersaturation gradients along faces decrease due to some equalization of the concentration by convective currents, and this makes it possible to prepare uniform crystals at considerably higher supersaturations (rates of growth) than under molecular diffusion conditions.

3. Forced Convection. This is produced by the action of external forces or by the forced motion of a crystal in solution.

There are no basic differences between forced and natural convection. When the velocity of motion of a liquid with respect to a crystal is increased, the thickness of the boundary layer decreases and the supply of matter to a face of the crystal increases. Therefore, by increasing the rate of motion of a solution, we can increase the rate of growth of the faces. However, we cannot continue this process indefinitely. The measurements of Cartier, Pindzola, and Bruins (1959) have shown that when the rate of motion is 10-30 cm/sec the rate of growth of single crystals reaches its maximum value and this value depends on the temperature and supersaturation of the solution. An even more important factor is that when the rate of motion of the solution is increased, the supersaturation gradients along the faces are reduced and this makes it possible to grow, relatively rapidly, large uniform crystals which would be difficult to produce under other conditions. There is a formula which can be used to estimate the conditions (size of a crystal, rate of motion of a solution, rate of growth) which ensure that uniform crystals are produced. This formula gives the rate of diffusion of matter v to a face which is parallel to a laminar current of the solution in which a crystal is situated [Levich (1959)]

$$v = 0.34 \frac{D^{2/3}}{\nu^{1/6}} \sqrt{\frac{U}{x}} \Delta C.$$

Here, D is the diffusion coefficient; ν is the kinematic viscosity of the solution; U is the velocity of the solution; x is the distance from the front edge of the face; ΔC is the supersaturation.

This formula was also deduced (independently of Levich) by Carlson (1958) but his formula has a somewhat different numerical coefficient.

Knowing the rate of growth of a face at a given supersaturation, we can use Levich's formula to calculate the distance from the front edge of a face, i.e., the crystal size, at which the rate of diffusion is insufficient for further growth so that the solution is depleted of the solute and a hollow appears on the surface. If we know the rate of growth and the size of a crystal we can calculate at what velocity of motion of the solution no inclusions will form in the crystal. Calculations show that the velocity of a solution ensuring the uniform growth of KH_2PO_4 and $NH_4H_2PO_4$ crystals is 9 cm/sec. This value is in satisfactory agreement with the experimental results of 10-15 cm/sec [Carlson (1958)].

Levich's equation is valid only for a smooth surface parallel to the current of a solution flowing past it. When a macroscopic step appears at the front edge, we may expect turbulence behind this step, i.e., the mixing of the solution, which increases the rate of diffusion but is ignored in Levich's treatment. If the faces are oriented as shown in Fig. 39, stationary eddies are formed above them. A solution captured by an eddy loses its solute as the crystal grows in size. Thus, the rate of growth of that part of the face below an eddy slows down. A hollow appears in the face and is followed by an inclusion (Pis'mennyi (1960)].

It has been reported that, in the same crystal, inclusions of a solution may be formed in growth pyramids on some faces and not others, although the growth rates of these faces are approximately equal [Petrov (1964)]. To explain this effect, it is necessary to take into account the state of a solution near the surface of a crystal. As demonstrated in §5, the surface of a crystal has an orienting influence on a thick layer of the solution. The thickness of such a layer may reach hundreds and thousands of angstroms. In such an oriented layer, the mobility of the molecules is lower than their mobility far from the crystal. Since the orienting influence of different faces is, in principle, different, the number of inclusions in the growth pyramids of different faces should be different in spite of the same growth rates of these faces. In the case of potassium nitrate, the rate of formation of inclusions at similar growth rates is in good agreement with the degree of similarity of the atomic structure of different faces of crystals of this compound and the faces of ice pinacoids. Inclusions are more frequent when the similarity of the structures is closer.

When the temperature is increased, the chemical binding becomes weaker. A consequence of this is a characteristic decrease in the thickness of the boundary layer (desolvation) at various faces, which reduces the frequency of formation of inclusions. A similar specific desolvation is found also when the supersaturation is increased [Kleber (1957)]. However, in the latter case, the concentration gradients increase more rapidly along the faces and, therefore, the number of inclusions increases.

The volume diffusion and the diffusion in an ordered layer of a solution near a crystal face may be affected by an impurity. For example, an appreciable increase of the rate of diffusion is observed when H^+ and OH^- ions are added to a solution. These ions have anomalously high mobilities [Samoilov (1957)] and, therefore, they tend to disorder the structure of a solution. The rate of growth of KNO_3 crystals increases in the presence of these ions (Petrov and Rozhnova (1963)]. Since potassium nitrate is not hydrolyzed and these ions (according to Paneth's rule) should be adsorbed appreciably on KNO_3, their effect may be explained by an increase in the rate of diffusion. In other cases, we must take into account the formation of strongly adsorbed compounds.

The symmetry of the diffusion field around a crystal depends on the growth conditions and on the characteristics of the relative motion of a crystal and a solution (the agitation, the motion of a solution on one side of a crystal, or some other type of motion). A crystal acquires a particular form which depends on the symmetry of the diffusion field; its external symmetry does not reflect the symmetry of its structure. According to a principle enunciated by Pierre Curie, a crystal retains only those elements of its intrinsic symmetry which are identical with the elements of symmetry of the medium. Therefore, knowing the symmetry of the medium, we can predict fairly reliably the shape of a crystal, the characteristic features of the morphology of its faces, the distribution of impurities, the sites of inclusions and other defects.

A detailed description of the relationship between the external symmetry of a crystal and the symmetry of the medium is given in Shafranovskii's book (1960).

CHAPTER II

METHODS FOR GROWING CRYSTALS

§7. Classification of Methods for Growing Crystals

Several hundreds of more or less different methods for growing crystals are known. The usual classifications of these methods are based on their technical features. Some workers thus distinguish the hydrothermal methods, the growth of crystals from solutions in the molten state, the growth from solutions at low temperatures, etc. In our opinion, the following features of crystal growth should be used as the base in the analysis of the similarities and differences between the various methods: (1) method of applying the driving force; (2) dependence of the driving force on time; (3) phase state and number of components in a medium.

1. **Method of Applying the Driving Force (the Difference of the Chemical Potentials).** The chemical potential is a function of temperature and pressure in one-component systems (melts of pure chemical substances or pure gases). In many-component systems (solutions), the chemical potential is also a function of the various concentrations. It is affected by the application of external fields, whether gravitational or electrical [Landau and Lifshits (1968)]. A change in any parameter of a system which includes a crystal and is initially in equilibrium, results generally in unequal changes in the chemical potentials of the crystal and the medium surrounding it. Therefore, the crystal either grows or dissolves. In view of this, we must consider the following parameters whose changes give rise to a nonequilibrium state in a system.

a) **Temperature T.** A change in temperature is used to produce supersaturation by the methods described in §§8, 9, 10, and 12.

b) **Pressure P.** A change in pressure is used to obtain new phases in the case of polymorphic transitions, such as the graphite-diamond transition. Since the solubility depends on pressure, a change in pressure may transfer a solution from an equilibrium to a nonequilibrium state.

c) **Concentration C**, which increases during a chemical reaction (cf. §11).

d) **Changes in the equilibrium concentration C_0** (due to a change in the composition of a solvent). For example, the addition of ethyl alcohol to a saturated aqueous solution of any electrolyte reduces its solubility (equilibrium concentration). The addition of alcohol results in the "salting out" of the electrolytes, which supersaturates the solution.

e) **Partial pressure of the solvent vapor above a solution, P_s.** The reduction of the vapor pressure of a solvent below its equilibrium value causes the evaporation of the solvent, which alters the concentration of the solution (§13).

f) **Electrical potential V**, causing chemical reactions, which may increase the concentration of a new substance above the equilibrium value. Here, we must include all electrochemical reactions which are used in the preparation of metal crystals.

g) **Gravitational potential** nG. Here, G is the force of gravity, equal to mg. The gravitational potential separates the components of a medium in accordance with their densities and this may produce regions in which the concentration of some component is higher than the equilibrium value. In the gravitational field of the earth $n = 1$. In centrifuges $n > 1$. Shternberg (1961) was the first to grow crystals in a centrifuge.

All these parameters are related in different ways to the chemical potentials of the medium and the crystal. Therefore, to establish and maintain the necessary constancy of the driving force we need appratus of great complexity. In view of this, although a particular crystal phase of a substance can be produced using several methods, one selects the method which is most convenient and matches best the aim of and the resources available to the experimenter. We shall discuss this in more detail in §22.

2. **Dependence of the Driving Force on Time.** Here, we can distinguish two possibilities:

a) The chemical potential varies from point to point in the crystallization medium but the nature of the potential distribution does not vary with time (steady-state growth conditions). The constant difference between the potentials in a system should remain unchanged with time because of the constancy of the difference between the potentials external to the system.

b) The chemical potential varies from point to point and with time (non-steady-state growth conditions).

The difference between these two cases is of basic importance. Under steady-state conditions, we can prepare crystals continuously. Moreover, steady-state conditions give us the best chance of obtaining very uniform crystals, whereas the variation of any parameter (temperature, concentration, etc.) alters the crystallization process and reduces its uniformity.

3. **Phase State and the Number of Components in a Medium.** A medium may be gaseous, liquid, or solid. The number of components may be one (pure substances) or many (solutions). These aspects are common to all methods and fundamental in the growth of crystals under hydrothermal conditions, from molten solutions, from solutions at low temperatures, etc.

Since a particular growth problem may have a number of technical solutions, we can distinguish the methods by the appratus used in growing crystals. The growth technique depends on the selected method and on the apparatus; its characteristics are: (a) the procedure used, the various operations involved, and their sequence; (b) the actual values of the crystallization parameters and the working conditions. These include the concentration of a substance, supersaturation, temperature, pressure, stirring conditions, etc.

We shall describe some modern laboratory methods for growing crystals from liquid many-component media and the techniques employed in these methods.

A. CRYSTALLIZATION UNDER STEADY-STATE CONDITIONS

§8. Crystallization under Thermal Convection Conditions

This method includes the widely employed techniques for growing crystals in a gaseous medium using "transport" reactions and practically all the modern hydrothermal techniques. The thermal convection method is currently employed to grow crystals from molten solutions. It is used also under low-temperature conditions [Walker and Buehler (1950); Petrov and Treivus (1960)].

The principle of the method is as follows. By maintaining different parts of a crystallizer at different temperatures, a temperature gradient is established, and convection is produced in a solution. A substance placed in the hotter region is dissolved and is transported by convection to the colder region. Here, the cooling makes the solution supersaturated and the excess solute is captured by a growing crystal, while the solvent returns to the hotter region. For a substance with a retrograde temperature dependence of its solubility, the dissolution and growth regions are interchanged, i.e., a substance is dissolved in the colder region and grows in the hotter zone.

Figure 17 shows a wide test tube whose lower part is immersed in a liquid thermostat kept at a temperature T_1. The upper part of the test tube is in an air thermostat or simply in air, where a lower temperature T_2 is maintained. In the absence of a substance in such a tube there is a constant convective exchange between the upper and lower parts due to the change in the density because of the temperature gradient (naturally, this convective exchange stops when the temperatures in the thermostats are altered so that $T_1 \leq T_2$, or when a liquid of very high viscosity is used).

During such convection, a steady temperature distribution is established (Fig. 18). The temperature difference between two points along a vertical line in the crystallizer tube increases when the distance between these points is increased, when the diameter of the crystallizer is reduced, and when the viscosity of the solution is increased.

We must mention particularly that the use of a liquid heat-transport agent (water) in the colder part of the tube produces, for the same temperature difference, a stronger supercooling and more rapid precipitation of the substance on the walls of the crystallizer tube than can be achieved by the use of a gaseous transport agent such as air. This is because the conditions for heat transfer through glass and water differ considerably from the conditions of heat transfer through glass and air. Therefore, if a liquid coolant is used, the temperature difference between the upper and lower parts of the tube should be much less.

In a vertical tube, the temperature at a given point is, on the average, constant but there are temporary deviations from the average value, which may reach 1 deg when the temperature in the lower part is about 50°C and the upper part (in air) is kept at about 20°C (this applies to a tube

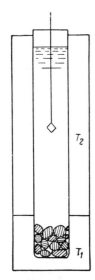

Fig. 17. Basic arrangement used in crystallization by thermal convection in a solution.

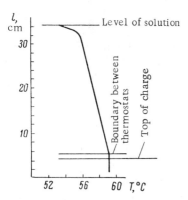

Fig. 18. Approximate temperature distribution along the height of a tube (tube diameter 30 mm; temperature in the heating thermostat 60°C; temperature in the cooling thermostat 25°C).

30 mm in diameter and 300 mm long). These temperature fluctuations are due to fluctuations in the convection currents and the mixing of the rising and descending currents. If a crystal is set in motion in a vertical tube this results in the mixing of the convection currents and a satisfactory equalization of the temperature of the medium around the crystal. The nature of the motion is ordered, particularly in the case of inclined or horizontal tubes. In the latter case, the boundary between the hotter and colder zones should naturally be vertical. However, the nonvertical positions of the tube are less convenient if it is intended to use dynamic crystallization conditions. Moreover, "stray" crystals become more easily attached to the inclined wall of a tube.*

When a substance which can be dissolved in a given liquid is placed at the bottom of a tube, the convection pattern becomes more complicated. The density of the liquid is now affected by the concentration of the solute, as well as by the temperature. If the rate of increase of the density due to the dissolution of the solute is greater than the rate of decrease of the density because of heating, the solution at the surface of the solute becomes more dense than the liquid above it. Consequently, the solution does not rise and no convective motion is obtained. The solution splits into layers. If the solute dissolves slowly or its solubility depends little on temperature, the concentration of the liquid next to the solute differs little from the concentration elsewhere in the solution and the heating is sufficient to reduce the density of the solution and set it in convective motion. To ensure convection, it is necessary either to use (instead of water) a solvent with a saturation temperature intermediate between the temperatures of the colder and hotter zones or to stir the solution. Moreover, it is necessary to ensure that the top of the solute (the charge) is 5-15 mm lower than the boundary between the hotter and colder zones. The descending solution is heated in the region 5-15 mm above the charge; it is also heated during its ascent after passing above the charge. If the surface of the charge is higher than the boundary between the hotter and colder zones the convection of the solution stops (in the case of substances which have a high solubility gradient), and the temperature in the crystallizer becomes almost equal to the temperature in the upper zone. If the top of the solute is much lower, two regions of convection are frequently formed, the exchange of matter between which is usually slow. The boundary between these regions is close to the boundary between the hotter and colder zones.

Let us now consider the actual technique used in growing crystals by this method.

---

*"Stray" crystals are those which appear spontaneously in a solution. They reduce the amount of the solute in a solution and slow down the rate of growth of the main crystal.

The crystallizers are usually test tubes 30-40 mm in diameter and 200-250 mm long. In the simplest case, the upper part of a tube is not placed in a thermostat. A laboratory stand supports the tube in the vertical position. The hotter part of the crystallizer, filled with a charge to a height of 30-40 mm from the bottom of the tube, is placed in a thermostat. If the solubility of the charge does not depend greatly on temperature, the solution in the crystallizer may be unsaturated at room temperature. If the solubility depends strongly on temperature, the solution should be saturated approximately at a temperature T_x defined by the formula

$$T_x = T_1 - \frac{T_1 - T_2}{4}.$$

This formula is based on the results of the measurements of the temperatures in tubes under normal experimental conditions. If a solution is prepared in advance and its concentration is equal to the concentration in the growth zone, spontaneous crystallization may begin in a tube and stray crystals may form on the walls of the tube and on the surface of the solution. Moreover, in solutions of substances whose solubility depends strongly on temperature, the following phenomenon may be observed when a slightly supersaturated solution is poured into a test tube. Crystallization begins in the hotter part of the tube and the charge placed in that part of the tube increases in volume. This reduces the heating area of the solution and, consequently, the temperature in the whole system. The fall in the temperature results in the precipitation of further amounts of the solute, which reduces the temperature still further, and so on, until the whole of the hotter zone becomes filled with the solute. Convection then stops and the temperature of the solution finally becomes equal to the temperature of the cold part of the tube. At this temperature, the solute precipitates completely. However, if a strongly unsaturated solution of a substance is poured into a tube and the solubility of this substance depends strongly on temperature, the whole of the substance may be dissolved in the hotter zone.

After the crystallizer has been set up, a stirrer is placed in it. The best method of stirring the solution during the preparatory and growth stages is to use a reciprocating motion, which improves the quality of a growing crystal (§16). The stirrer is usually a glass rod with several spherical projections of larger diameter along its length; during its motion the stirrer should not disturb the charge at the bottom of the crystallizer. The level of the charge is watched during the stirring. If the level of the charge changes appreciably because of dissolution, it is necessary to add a further batch of coarse crystals. The stirring is continued for several days. Next, the stirrer is taken away and the solution is rested for 2-3 h in order to establish normal temperature conditions. After this time, a crystal seed is introduced into the solution at a depth of 30-40 mm. If this seed dissolves (cf. §27) it means that the solution is not yet thoroughly mixed and the stirring should be continued. Sometimes it is found that the first seed grows faster than the other seeds and, therefore, the quality of the first crystal is lower than that of the other crystals. This is because the stirring raises the temperature of the solution and its concentration above their values in the case of free convection undisturbed by forced mixing. In such a case, the first crystal forms in a region with an excess of the solute. Therefore, it is necessary not to attempt to correct the conditions until a second crystal is produced. If the second crystal is also defective, it is necessary to take certain measures, which depend on the nature of the defects observed in the first two crystals (cf. §21).

The optimum conditions cannot be determined easily. This is because there have been very few systematic investigations in which the solubility and the temperature coefficient of the solubility have been compared with the parameters of the crystallization process. Moreover, there have been very few thermal studies of free convection in solutions in the range of values of the density, viscosity, and geometrical dimensions of interest in crystal growth processes [Ostroumov (1952); Slavnova (1963); Smid and Sommer (1969)]. Therefore, it is necessary to use the general relationships governing convection processes and to select experimentally the optimum crystallization conditions by trial and error.

The usual difficulty in such experiments is the formation of a fine-grained crust on the surface of a solution, which is the result of strong supercooling of the surface layers and, possibly, the evaporation of the solvent. In some cases, it is sufficient to cover the surface of a solution with a layer of liquid petrolatum, 20-30 mm thick (the thickness of such a layer should be greater than the amplitude of vibrations of the crystal-holder or seed). This layer reduces somewhat the supercooling of the solution and prevents the evaporation of the solvent. A wet seed can be introduced quite safely through such a layer into the solution. Naturally, in the case of solutions containing organic substances liquid petrolatum should be used only after checking that there is no reaction between it and the solution. If the formation of stray crystals on the surface of a solution is still observed, it is necessary to reduce the difference between the temperatures of the colder and hotter zones, or to reduce the height of the solution in the tube by removing some of it with a pipette. The same procedure should be followed if stray crystals appear on the walls of the test tube.

After a change of the temperature of the solution and after the removal of some of the solution by a pipette, time should be allowed for the solution in the test tube to reach equilibrium again, perhaps by mixing and lowering the tube into the hotter zone if the change in the temperature has caused considerable precipitation of the solute at the bottom of the test tube. Usually, the temperature in the colder zone is within the range 25-50°C and the temperature in the hotter zone is 40-80°C, so that the external temperature difference between the zones is within the range 10-40 deg C. The internal temperature drop under such conditions may be only several degrees. Under these conditions, good-quality crystals can be prepared of substances such as lead, strontium, and barium nitrates, sodium bromate, etc.

The rate of growth of a crystal under thermal convection conditions is approximately proportional to the product of the rate of convection v and the temperature difference ΔT within a test tube (the internal temperature drop is a quantity which is governed primarily by the supersaturation of a solution near a crystal). However, the supersaturation of a solution is always lower than that calculated from the difference of temperatures of a charge and a growing crystal. The main reason for this is that the solution flowing past the charge does not have sufficient time to reach a state of equilibrium with the charge. Obviously, the rate of dissolution depends strongly on the area of the charge.

The product of the temperature difference and the rate of convection represents approximately the amount of substance which is in excess of the saturation concentration and which moves past a growing crystal. There are ways in which ΔT and v can be altered: we can alter the temperature difference between the two thermostats (the external temperature difference); we can change the distance between the growing crystal and the charge; and we can alter the diameter of the tube. The rate of convection varies from point to point and its "effective" value near a crystal is difficult to determine. It is possible to determine only approximately its order of magnitude from the motion of small particles in a solution. The velocity of such motion is within 20-80 cm/min. An increase in the distance between a growing crystal and a charge reduces the rate of convection near the crystal. This is because, under laminar convection conditions, part of the ascending current turns back and descends so that the greater the height of the tube, the slower is the motion of the convection current. A reduction of the rate of convection near a growing crystal increases the temperature drop between this crystal and the zone where dissolution takes place. Thus, both quantities (ΔT and v) are altered in inverse proportion. However, the rate of growth is proportional to the product of these quantities. Therefore, the rate of growth is maximum at some optimum distance between a growing crystal and the charge (Fig. 19).

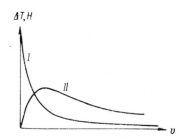

Fig. 19. Temperature difference ΔT (I) and amount of the solute H (II) passing past a growing crystal, as a function of the rate of convection. Curve II is plotted by multiplying together the coordinates of the points of curve I. The rate of growth is approximately represented by curve II and has a definite maximum.

A reduction of the diameter of the crystallization tube increases ΔT and reduces the rate of convection, i.e., the rate of growth once again should have a maximum at some optimum diameter of the tube. For example, in the growth of quartz crystals, the diameter of the autoclave is not altered but a horizontal partition pierced by apertures is placed in such a way as to spread the growth and solution zones [Walker and Buehler (1950)]. The apertures in the partition represent only 10% of the total cross-sectional area of the autoclave. In the method described here, the use of such a partition is undesirable because stray crystals frequently appear on the partition. Instead, we can use tubes with constrictions between the solution and crystallization zones. To prevent the formation of stray crystals, a constriction should be fairly gradual.

During the growth of a crystal, the parameters of the process change somewhat. This is mainly due to the gradual transfer of the charge from the dissolution zone. The charge is usually dissolved in such a way that a mushroom-like projection is formed on its surface. Consequently, the temperature in the solution increases. Generally, the supersaturation varies with time because of the increase in the surface of a growing crystal, the change in the area of the surface of the dissolving charge, and the change in the temperature in the tube. When fairly large (3-4 cm^3) crystals are needed, it is desirable gradually to raise the tube as the charge is used up. To avoid sudden changes in the supersaturation, the tube should not be raised by more than 2-3 mm each time.

To grow larger crystals of a substance which has a strong temperature dependence of the solubility, we should use the apparatus shown in Fig. 34. In this case, a charge forms a layer 1-2 cm thick in a crystallizer. The height of the liquid column is 15-20 cm. The solution is heated by circulating water in a thermostat. The upper part of the tube, as well as the surface of the solution, are cooled by air trapped between the solution and the cover of the crystallizer. By varying the level and temperature of the water in the thermostat, we can control the internal temperature drop between the bottom and top parts of the solution. For example, in growing potassium nitrate crystals, the best conditions are obtained when the level of water in the thermostat and the level of the solution are the same (the temperature in the thermostat is kept at 35-37°C). In these experiments, cooling takes place only at the surface of the solution. Under such conditions, the internal temperature drop is not more than one-tenth of a degree.

If the level of the water in the thermostat is close to the level of the charge in the crystallizer, the temperature conditions are unstable. However, if the crystallizer is immersed in the thermostat to about half its height, the temperature fluctuations are about 10 times smaller in the crystallizer than in the surrounding air.

§9. Crystallization under Concentration Convection Conditions

In this method, as in the thermal convection technique, the driving force in the crystallization process is the temperature difference. However, the motion of the solution (which is again due to gravitational forces) stems from the difference between the densities resulting from the difference between the concentrations in the solution. The method can be used for substances whose saturated solution density increases when the temperature is increased.

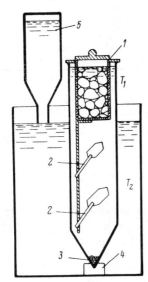

Fig. 20. Apparatus used to grow crystals by the concentration convection method. 1) Charge chamber containing the substance which is being dissolved; 2) crystal holder; 3) stray crystals; 4) rubber support; 5) device for maintaining a constant water level in the thermostat.

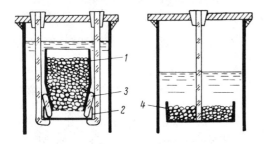

Fig. 21. Two variants of the construction of the charge chamber. 1) Porcelain crucible with a porous bottom or filter with a porous glass plate; 2) tray used to prevent stray crystal formation; 3) glass rods for supporting the crucible or filter; 4) glass tray attached to a glass rod.

The method, in which the concentration currents are used to transport matter from a hotter to a colder zone, was employed first by Spezia [Buckley (1951)] to grow quartz crystals under hydrothermal conditions. Later, a convenient method for growing crystals from aqueous solutions at room temperature was described by Belyustin (1961, 1965). In contrast to the thermal convection method, a charge which is being dissolved is placed above a growing crystal.

The crystallizer is usually in the form of a tube 150-300 mm long and 40-50 mm in diameter (Fig. 20). The lower part of the tube tapers down to a cone in order to impede the growth and accumulation of stray crystals. The crystallizer can also be in the form of a cylinder or a tall beaker. The chamber containing the charge is placed in the upper part of the crystallizer. This chamber is usually made of a transparent plastic. The diameter of the chamber should be 5-6 mm smaller than the diameter of the crystallizer. The walls of this chamber have apertures of 1-2 mm inclined downward into the chamber in order to prevent the washing out of the powdered charge during the loading of the crystallizer and to impede the formation of stray crystals. The chamber can also be a porcelain crucible with a porous bottom or a glass filter (type No. 1 or 2). In this case, it is necessary to place a tray under the chamber in order to prevent stray-crystal formation (Fig. 21). The charge chamber can also be in the form of a small crystallizer with low walls. Seeds are fixed as shown in Fig. 22. The temperature difference between the charge chamber and the crystallization zone may be established by immersing the crystallizer in a vessel containing water. Evaporation reduces somewhat the temperature of the water compared with the surrounding air (by 0.1-0.2 deg C). This difference is quite sufficient for the growth of crystals. The large mass of water also stabilizes the temperature in the crystallizer.

The sequence of the operations is as follows. A solution is prepared in advance (the saturation temperature of the solution should be equal to the crystallization temperature). The charge chamber is filled with the required substance and then the chamber is washed in a solution of the same substance in order to remove dust particles. The best results are obtained by placing fairly large crystals in the charge chamber. The solution is heated to a temperature 1.5-2.5 deg C higher than the saturation temperature and is poured into the crystallizer. If the solution is heated too much its concentration may change considerably because of the

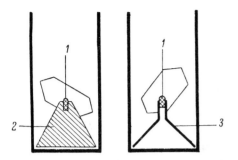

Fig. 22. Two methods for the attachment of the seeds used in beakers and cylinders. 1) Seed; 2) cone made of plastic; 3) glass funnel.

more rapid dissolution of the substance placed in the charge chamber. An insufficiently high temperature increases the chances of formation of stray crystals. The charge chamber is then immersed slowly in the crystallizer. Depending on the nature of the holder, a seed is either placed in the crystallizer before the solution is poured in, or immersed in the solution together with the charge chamber. Next, the crystallizer is hermetically sealed and placed in a thermostat. This must be done carefully in order not to shake the liquid in the crystallizer, which would give rise to stray crystals. The crystallizer should not be touched before the extraction of the grown crystal, and the crystallizer should be placed on a rigid vibration-free mounting.

After the extraction of the grown crystal, the whole procedure is repeated after the solution has been poured into a separate flask and heated to the required temperature. The routine is acquired rapidly and mishaps are rare after three to five runs. The supersaturation and rate of growth can be controlled in several ways. Firstly, the larger the distance between the charge chamber and the growing crystal, the higher is the rate of growth. Secondly, the smaller the number of seeds in the tube, the higher is the rate of growth. The rate of growth can also be accelerated by an additional temperature difference between the charge chamber and the growth zone; this can be done by heating the tube surrounding the charge chamber and thus increasing the rate of solution. The temperature in the charge chamber can also be increased by a heater wound around the outer tube and supplied with a stabilized current. The very small power supplied to the heater is selected by trial and error; it is usually sufficient to increase the temperature by a fraction of a degree.

Instead of placing the crystallizer in a water bath, one can cool the tube at the level of the growing crystal by winding wet gauze around it and placing the end of the gauze into a beaker containing distilled water. The gauze collar is made up of three or four layers about 2.5 cm wide. The width of the gauze collar, its distance from the water container, and the density of the wick feeding the gauze collar are selected by trial and error. If the gauze collar does not have the right properties the water will evaporate before reaching the collar. Crystals up to 3 cm in diameter grow in two weeks to three months, depending on the substance. The higher the solubility of the substance and the greater the difference between its specific gravity and that of the solvent, the more rapid is the growth of the crystals. A constant temperature must be maintained and it must be remembered that relatively large but gradual fluctuations of temperature are less harmful than shorter but rapid fluctuations. This is because the rate of growth of a crystal in the charge–solution–growing-crystal system is, to some extent, controlled automatically. When the temperature is increased the rate of solution increases (in the case of substances with a positive temperature dependence of the solubility) and the concentration of the solution becomes higher. However, since any increase in the temperature applies also to the solution surrounding a crystal, it follows that, in spite of the increase in the concentration near the crystal, the supersaturation remains practically constant and the rate of growth does not change very much. Similar effects are observed when the temperature is reduced. This can be seen by examining the solubility diagram (Fig. 6). However, such automatic stabilization is not sufficient and, therefore, active thermostating is required (cf. §15). When active thermostating is used, the method can be employed to grow crystals of substances at very low supersaturations. In the mass-production of crystals under identical conditions, it is more convenient to use one thermostat for several crystallizers.

This method of crystal growth is particularly convenient in research when an investigator requires only small crystals, when the amount of the substance under investigation is small, or when it is expensive. In this respect, the more widely used methods described in the subsequent sections (by the evaporation of the solvent or by varying the temperature) are usually less suitable.

§10. Crystallization Under Forced Convection Conditions

In this method, the driving force is, as in the preceding two methods, the temperature difference between the zone in which the solution becomes saturated and the zone in which a new crystal grows. However, in contrast to the preceding two methods, the transport of matter in the solution from the saturation zone to the growth zone is forced by pumping. The ordered motion in the medium allows the use of three thermostats. This makes it possible to control the mass transport from the dissolution zone to a growing crystal, and the heat transfer in the system. The complexity of the apparatus used to exploit in full the possibilities of this method is fully justified by the accuracy and convenience of control and maintenance of the supersaturation level, temperature, composition of the crystallization medium, and the rate of renewal of the solution near a growing crystal. The latter facility is particularly important in crystallization processes taking a long time, and in continuous processes which can be realized by this method. The first variant of the apparatus for growing crystals by this method was described by Kruger and Fincke in 1910 [Buckley (1951)]. Buckley (1951) and Wilke (1963) describe many other variants of the method. Houghton (1965) has successfully grown large crystals of many substances by this method. The basic parts of the apparatus used in growing crystals by the forced convection methods are shown schematically in Fig. 23.

One of the present authors [Petrov (1962)] has suggested a method for the continuous growing of crystals with a given cross section. Petrov's apparatus is shown schematically in Fig. 24. The crystallization chamber (Fig. 25) is intended for growing a single crystal, and this crystal can grow along one direction, i.e., along one face. The spreading of the face is limited by the walls of the chamber. A stream of supersaturated solution is supplied directly to this growing face.

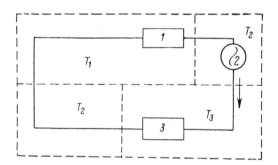

Fig. 23. Schematic diagram of the apparatus for growing crystals by the forced convection method. 1) Crystallization chamber; 2) pump; 3) feed chamber; T_1) thermostat for supercooling the solution in the crystallization chamber (enclosed by dashed lines); T_2) thermostat ensuring that the solution is heated above the saturation temperature; T_3) thermostat maintaining the saturation temperature in the feed chamber.

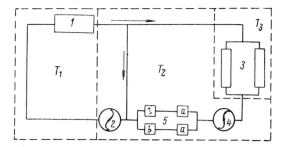

Fig. 24. Schematic diagram of the apparatus for growing crystals from solutions. 1) Crystallization chamber; 2) pump ensuring the fast flow of the solution past a crystal; 3) solution (feed) chamber; 4) pump which circulates the solution through the feed and purification units; 5) system of filters (a) and chemical absorbers (b). T_1, T_2, and T_3 have the same meaning as in Fig. 23.

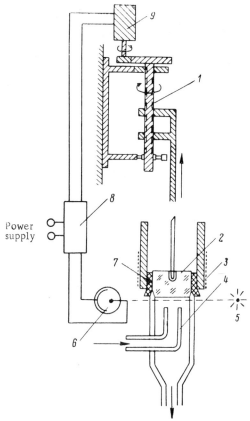

Fig. 25. Chamber for the production of cylindrical crystals of small diameter with an automatic device for the pulling of a crystal. 1) Pulling mechanism; 2) crystal; 3) external heater of the crystallization chamber; 4) nozzle supplying the solution; 5) source of light; 6) photocell (photodiode); 7) elastic spacer; 8) amplifier for pulses generated by the photocell and passed onto the motor (9).

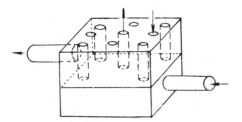

Fig. 26. Lower part of the crystallization chamber with several nozzles for the supply of the solution to a growing crystal.

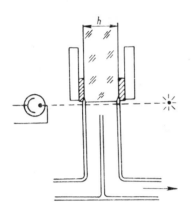

Fig. 27. Crystallization chamber for the preparation of crystals in the form of plates; h is the thickness of a plate.

To prepare a crystal of 10-15 mm diameter, it is sufficient to use a single nozzle of 6-10 mm diameter. For crystals of larger diameters or more complex cross sections, several nozzles are required which can be distributed as shown in Fig. 26. The maximum distance between two nozzles can be calculated using the formula given in §6. To prepare crystals in the form of plates, one can use the chamber shown in cross section in Fig. 27.

As the face grows, it approaches the nozzle and may close it if the crystal is not moved away. The crystal is extracted gradually from the chamber by an electric motor controlled by a relay operated by a photoelectric scanner (Fig. 25). The device works by switching on a mechanism which raises the crystal 1 at the moment when a beam of light from a source 5 is interrupted; this beam passes through the growth chamber to a photocell 6. A satisfactorily constructed photoelectric system makes it possible to keep the growing surface of a crystal at a constant distance from the nozzles (to within 1 mm). Such accuracy is sufficient for nozzles located 4-6 mm from the surface of the growing crystal.

The mechanism which raises the crystal should be connected rigidly to the crystal and the direction of the force applied to raise the crystal should coincide with the axis of the chamber. It is necessary to heat the side walls of the chamber 3, otherwise the crystal can jam against an elastic spacer 7 and a large force might be required to extract the crystal from the chamber. When the solution is supplied at a high rate and its specific heat is high, the crystallization process is not disturbed if the upper part of the crystal is kept at a temperature 1-2 deg C higher than the saturation temperature of the solution.

The concentration of the solution which flows past the face of the growing crystal at a high velocity remains practically constant. In principle, the solution could be passed through the next crystallization chamber but after contact with the growing crystal (particularly if the solution is not carefully freed of mechanical impurities) the possibility of stray-crystal formation is greatly increased. Therefore, the solution is passed from the crystallization chamber directly into heat exchangers in which it is heated to a temperature 3-4 deg C higher than the saturation temperature (Fig. 24, a zone denoted by T_2). After this stage, the solution is pumped to a zone kept at a temperature T_1, where it is supercooled and directed back to the crystallization chamber. The solution is circulated by a centrifugal pump which moves the solution past the growing crystal face at a velocity of 0.20-0.25 m/sec.

To ensure a constant supply of the material being dissolved and to purify the solution, the solution is passed through a second circuit in which the temperature is not lower than the saturation temperature. In this circuit, the solution is circulated by a powerful pump 4. A membrane or plunger-type pump is suitable for this purpose. The chambers in which the original substance is dissolved are placed in a zone where the saturation temperature T_3 is maintained. For continuous operation, it is useful to have two such chambers: while one is in use, the other may be cleaned or loaded. The actual construction of these chambers is not very important provided the substance is in sufficiently good contact with the flowing solution and the solution emerging from the chamber is not too turbid. After the solution has emerged from the solution chamber, its temperature is raised and it is passed through a cleaning system 5. This system has filters (a) which remove suspended particles from the solution. The areas of these filters and their porosity should ensure the required purity of the solution but they should not raise the pressure too much between the pumps and the filters.

In order to reduce the pressure, which increases as the filters block up, it is advisable to use substance from which mechanical impurities have been removed before use. To measure the degree of block-up of the filters and the resultant pressure rise in the system, it is advisable to insert a manometer in front of the filters which will signal the need for the replacement of the filters. A double purification system makes it possible to change the filters without stopping the flow of the solution.

A solution must be purified chemically to remove any impurities which might affect the growth of crystals; these include products of the decomposition of the solution or of the leaching out of the walls of the system, or impurities present in the original reagents. This can be done in two ways. The simplest method is a continuous replacement of the solvent by providing an outlet for the spent solvent (which must be removed slowly so as not to disturb the conditions of crystallization) and an inlet for the admission of a fresh purified solvent. For example, if it is known that a harmful concentration of impurities is accumulated in two weeks, it means that the whole of the solution in the crystallizer must be replaced gradually during this period. If this is done, the impurity concentration remains constant and its value does not exceed half the critical value at which harmful effects begin.

A less convenient method is to purify a solution chemically within the system itself. Chemical absorbers (Fig. 24, unit 5b) are essential in the case of unstable compounds when the accumulation of decomposition products in the solution is so rapid that the replacement of the solution is inconvenient. Naturally, both methods can be used simultaneously. After purification and sufficiently long heating, to achieve the necessary temperature, the solution is admitted again into the crystallization chamber.

In such a system, the control of the various parameters governing the crystallization conditions does not present any great difficulty. It is more difficult to decide whether such control is necessary because the lower part of a crystal, located in the nontransparent part of the growth chamber, cannot be seen at all and the part of the crystal outside the elastic spacer (placed between the crystal and the walls of the chamber) is opaque because it contains many fine surface inclusions. Moreover, the growth is complete in this part of the crystal and, therefore, examination can only indicate whether the conditions have been correct during its growth. One can also control the conditions by observing the growing surface of a crystal; this is not a convenient method but it can be used if necessary. An MBS-2 binocular microscope can be used for the examination of the surface. The light source and the microscope are arranged in Fig. 28.

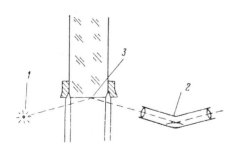

Fig. 28. Location of a light source (1) and a microscope (2) for the observation of a growing face of a crystal (3).

The apparatus and method described here can be used to grow substances of relatively high solubility, such as Rochelle salt, ammonium or potassium dihydrogen phosphate. For these two phosphates, which are sensitive to heavy cation impurities (calcium and iron), the best conditions are obtained by using a chemical purification unit.

The use of such apparatus is justified only when large numbers of crystals of one substance (but of different shapes) are required. To increase the efficiency, we can connect several crystallization chambers in parallel. In view of the considerable complexity of such apparatus, the method just described is used only rarely under laboratory conditions. However, this method may become an important industrial method for the growth of crystals from solutions. It has the following advantages: 1) the constancy of all the crystallization parameters; 2) the preparation of single crystals of required shape in a semi-finished form, which means that little of the material grown is lost in the final finishing processes; 3) the preparation of crystals can be made fully automatic.

The control of the crystallization parameters and the possibility of continuous observation (for example, using a cine camera) of a growing crystal make this method attractive in investigations of the crystallization from low-temperature solutions [Bliznakov and Kirkova (1957); Cartier, Pindzola, and Bruins (1959)].

§11. Crystallization by Chemical Reactions under Counter-Diffusion Conditions

In this method, an excess of a substance above the equilibrium concentration is produced by a chemical reaction. It can be used when the chemical reaction yields a substance whose solubility or equilibrium vapor pressure is less than that of the reacting substances. Among such reactions are

$$CaSO_4 + 2KF = \downarrow CaF_2 + K_2SO_4;$$
$$Pb(C_2H_3O_2)_2 + Zn = \downarrow Pb + Zn(C_2H_3O_2)_2;$$
$$NH_3 + HCl = \downarrow NH_4Cl.$$

The reactions used for this purpose are usually fast and therefore measures must be taken to prevent too rapid supersaturation of the solution and the formation of fine-grained or colloidal materials.

It is known that the amount of a substance formed per unit time in such reactions is, in practice, governed solely by the rate of supply of reagents to the reaction zone, i.e., by the rate of their mixing. Therefore, the reaction can be controlled by varying the supply of the reagents to the reaction zone. To obtain orderly transport conditions and to make it easier to control the rate of reaction, it is usual to employ tubes filled with a solvent. The reagents are placed at the opposite ends of such a tube. For example, we can place AX and BY at the two ends of a tube (here, A and B are cations and X and Y are anions). These two substances are dissolved, diffuse toward one another, and react to produce a less soluble substance AY. Steady-state conditions in such a system are ensured by the formation of solutions saturated with AX and BY at the two ends of the tube and by the constant removal of the freshly formed substance AY from the solution. We must remember also another factor. At the beginning of the reaction, the concentration of the second reaction product (BX) in the solution will begin to increase. Therefore, initially the process will not take place under steady-state conditions. The increase in the amount of the substance BX will be limited by the formation of a solution which is supersaturated with respect to this substance. Such a solution eventually becomes labile and BX begins to crystallize. When the solution reaches a metastable state in respect to BX, the rate of diffusion of AX and BY, as well as the rate of formation of AY and of BX, reach a steady state. This steady state is maintained until one of the original substances is exhausted. However, if the solvent is continuously supplied with the original reagents, the system reaches steady-state conditions more rapidly, and we can prevent the undesirable appearance of a second phase.

The question is how to control the rate of crystallization in this method.

The formula which describes molecular diffusion (cf. §6) shows that the rate of supply of substances to the reaction zone can be controlled by varying the concentration gradient dC/dx. Under steady-state conditions, this gradient is given by

$$\frac{dC}{dx} \simeq \frac{C - C'}{l}.$$

Here, l is the distance from either of the original substances to growing crystals; C is the concentration of A ions in the solution near the substance AX (here, the solution is practically saturated); C' is the concentration of A ions in the solution near AY crystals formed by such a reaction. Usually, C' is negligibly small compared with C. If the solubility of AY is low, we can drop C' from the above expression and obtain approximately

$$\frac{dC}{dx} \simeq \frac{C}{l}.$$

Since dM/dt is directly proportional to dC/dx and the cross-sectional area (§6), it follows that the rate of supply of A ions to the reaction zone may be reduced by making the crystallization chamber longer or by reducing its diameter, or by using a less soluble original substance.

If the original substances have very different solubilities, the freshly formed crystals of AY may be deposited directly on the surface of the less soluble of the two original reagents. Thus, when zinc reacts with lead acetate, lead crystals grow on the surface of the zinc. When calcite ($CaCO_3$) reacts with cupric sulfate, both malachite and gypsum are formed. We have already mentioned the solubility as the decisive factor for the location of the area where the diffusing substances meet because the differences between the diffusion coefficients of different substances are, in general, within the same order of magnitude, while the solubilities can differ very considerably. Sometimes, the growth of new crystals on the surface of an original substance is permissible but, in general, it is undesirable. It is frequently difficult to separate the newly formed crystals from the original substance and, which is more important, the newly formed crystals prevent the solvent from reaching the original reagent and this slows down the reaction and eventually stops it altogether.

Since the reaction takes place in a relatively large volume of the solution, we must ensure also that the freshly formed substance is transported to the surface of a growing crystal. When the rate of reaction is higher than the rate of crystallization in the growth zone, the freshly formed substance accumulates in the solution. Such accumulation accelerates somewhat the supply of the freshly formed substance to a growing AY crystal, near which the concentration is close to the saturation level; this accelerates somewhat the crystallization rate. However, if the accelerated rate of crystallization is insufficient to prevent continuous buildup of the freshly formed substance, the solution far from the growing crystal will sooner or later reach the labile state and new nuclei will form simultaneously. Fortunately, solutions of low-solubility compounds have usually a very wide metastable region. Thus, Morse and Pierce (1903) have shown that the concentration of a supersaturated solution may be 150 times higher than the saturation concentration without the transition to the labile state. Therefore, generally, there is no need to take measures to ensure that the reaction takes place in the direct

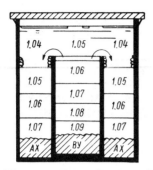

Fig. 29. Distribution of densities in the solutions above the original substances.

vicinity of a growing crystal. Moreover, spontaneous nuclei are surrounded by fairly wide zones, in which the formation of new crystallization centers is not very likely because of the transition of the solution to the metastable state.

In this method, seed crystals are not normally used. Spontaneously produced nuclei in a medium of low viscosity, insufficient to support such nuclei in the suspended state, become attached to the walls of the crystallizer or drop to the bottom of it. The cross section of the crystallizer in the case of spontaneous nucleation should be selected to ensure that the dimensions of the crystals in a given part of the tube differ little in size when they are nucleated during the same stage. However, if the sizes of the nuclei differ considerably, this suggests that fresh nuclei are being formed continuously, i.e., that the cross section of the reaction zone of the crystallizer is too large.

The appearance of a fine-grained material may also be due to macroscopic motion in the solution, arising from the difference between the densities of the solution in different parts of the crystallizer, or from the vibration of the crystallizer. It is known that the low velocity in molecular diffusion gives rise to strong vertical concentration gradients above the feed substance. Such gradients are particularly large in the case of substances of high solubility. The results obtained in such a case can be seen in Fig. 29.

This figure shows a smaller beaker placed within a larger one. At the bottom of the larger beaker, we have a substance AX and at the bottom of the smaller one we have BY. The solution of these substances and the upward diffusion produce a distribution of concentrations indicated in Fig. 29 by numbers representing the distribution of densities. When solutions of different densities meet at the edge of the inner beaker, the denser solution flows into the outer beaker (this is shown by arrows). Rapid mixing produces fine-grained crystals on the outer wall of the inner beaker and on the wall of the outer beaker. Such interaction between moving solutions also takes place in crystallizers with horizontal reaction chambers. The only effective way to prevent the formation of such fine-grained crystals is the use of media sufficiently viscous to prevent the mixing of solutions because of the differences between their densities. Substances used to increase the viscosity of solutions are sodium silicate ("liquid glass"), agar−agar, and gelatine [Henisch, Dennis, and Hanoka (1955)]. Henisch et al. have shown that in high-viscosity solutions we can obtain 10 mm crystals of low-solubility compounds, such as calcium tungstate, lead iodide, and lithium fluoride. The disadvantage of this method is the possible effect of the additive (used to increase the viscosity of the solution) on the quality and shape of the growing crystals. In practice, this is not a serious problem because subsequent examination shows whether the crystals obtained in this way are of sufficiently good quality.

The technique and apparatus used in this method are extremely simple. We shall describe a variant which represents a slight modification of the equipment described by Henisch et al.

We can use ordinary laboratory beakers of two different sizes (Fig. 29). A hot solution of, for example, gelatine is poured into the inner and outer beakers to a height of 1-1.5 cm. Then, stoichiometric amounts of the original reagents are poured in. The solution is allowed to gel and then it is covered by a heated solution of gelatine which fills both the inner and outer beakers. The crystallizer is hermetically sealed to prevent evaporation. To seal the crystallizer, it is sufficient to cover it with a sheet of glass and close the gaps with plasticine. Usually, because the process is long and the diffusion coefficient depends weakly on temperature, such a crystallizer is not placed in a thermostat. We can also use a U-shaped tube (Fig.

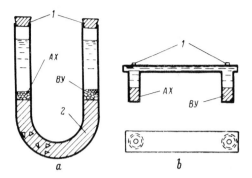

Fig. 30. Crystallizers used in growing crystals by chemical reaction: 1) rubber stoppers; 2) sodium silicate ("liquid glass"), solution of agar–agar or gelatine.

30a). In this case, the tube is filled to half its height with a solution containing the substance which makes it viscous, and then, after the solution has gelled, the solvent and the two substances are placed in the two arms of the U-tube. The gel should be sufficiently viscous for the crystals of the original substances to float on its surface without sinking. A convenient form of such a crystallizer is shown in Fig. 30b. In this case, the crystallization process can be observed with a binocular microscope. When gels are used in this process, the shaking or moving of the crystallizer is not harmful.

The preparation of crystals by this method is a relatively slow process: it takes weeks or months. The dimensions of the crystals produced in this way usually range from fractions of a millimeter to 10 mm. It may be possible to use this method to prepare crystals up to several centimeters in diameter.

The preparation of crystals of Zn, Pb, and Cd hydroxides, several millimeters in size, has been reported by de Haan (1963). This method has also been used by Brauns (1904), Chirvinskii (1904-1906), Berg (1957), and Wilke (1963).

Kašpar, Barta, and Nigrinova (1968) regulated the rate of diffusion of components in a solution by applying an electrical potential difference.

The advantages of the method include the simplicity of the apparatus and of the technique, and the possibility of preparing crystals of substances which can be grown only with very great difficulty under other conditions. The disadvantages include the slowness of the process and the small size of the crystals; however, the size is sufficient in many investigations.

B. CRYSTALLIZATION UNDER NON-STEADY-STATE CONDITIONS

§12. Crystallization by Cooling of Solutions

In this method the supersaturation is produced by a change in temperature usually throughout the whole crystallizer.

The solubility of most substances increases with temperature and therefore it is almost always necessary to cool a solution. There are, however, rare cases of substances whose solubility decreases when the temperature is increased and whose temperature coefficient of the solubility is small. Therefore, this method cannot be used to grow crystals of such substances (for example, $Li_2SO_4 \cdot H_2O$, $CaSO_4 \cdot 2H_2O$).

The method is usually called crystallization by cooling. The crystallization process is carried out in such a way that the point on the temperature dependence of the composition moves into the metastable solution region along the saturation curve in the direction of lower solubility. Since the volume of the crystallizer is finite and the amount of substance placed in it is limited, we cannot achieve simultaneously a constant temperature and a constant supersaturation. Therefore, the supersaturation requires systematic cooling. However, we know that a change in any crystallization parameter (such as temperature) unavoidably affects the crystallization process and gives rise to inhomogeneities of the structure and composition of a crystal. This is the main disadvantage of the cooling method.

In this method we can use a thermostated crystallizer, shown in Fig. 34. The volume of the crystallizer is selected on the basis of the desired size of a crystal and the temperature dependence of the solubility of the substance. The temperature at which such crystallization can begin is usually within the range 50-70°C. At higher temperatures, the technique becomes very difficult. It is necessary to heat the filter flask and the filter; moreover, when air is pumped out of the flask the solution begins to boil and the solvent is lost rapidly.

One must be aware of the possible harmful effects of the opening of a crystallizer during the manipulation of a seed. Moreover, when the temperature is increased the width of the metastable zone decreases and this means that the temperature must be maintained more rigorously and a higher standard of cleanliness must be observed. These difficulties can be overcome but they must be remembered if experiments are to be carried out at high temperatures. If the difficulties are too great and there are no special requirements as far as the crystal is concerned, we can use the thermal convection method without a change in the technique (cf. §8). Usually, the lower limit of cooling is the room temperature. However, to increase the efficiency of the method and to extract the maximum amount of the crystallizing substances from the solution it is recommended to cool the solution below room temperature. This can be done by circulating tap or iced water. In this way the temperature can be reduced by 10-15 deg C below room temperature.

We shall describe the techniques used in this method. A solution is prepared so that its concentration is saturated at a selected temperature which may be, for example, 50°C (cf. §24). Next, the solution is heated to 10 deg C above the saturation temperature, it is filtered, and immediately poured into a carefully washed crystallizer, which is kept in a thermostat maintained at 55-56°C. The crystallizer is closed with a temporary cover. The water level in the thermostat is made 10-15 mm lower than the level of the solution in the crystallizer. Under these conditions the solvent condensed on the walls flows back into the solution and dilutes somewhat the surface layer. Consequently, stray crystals are not formed at the surface of the solution.

Errors in the published data on the saturation temperatures, as well as errors in weighing and measuring the amount of the solvent, the loss of the solvent during heating and filtration, are all factors which may alter the saturation temperature of the solution at the moment of pouring into the crystallizer from the expected value. If the solution is strongly unsaturated, a seed placed in it dissolves rapidly and the large thermal inertia of the thermostat and the crystallizer prevents rapid establishment of the necessary saturation temperature which would prevent the dissolution of the seed. If the solution is supersaturated, the opening of the cover of the crystallizer and, particularly, the introduction of a seed may contaminate the solution and disturb the process.

To grow satisfactory crystals of many substances, supercooling amounting to several tenths of a degree is required. An increase in the supersaturation produces poor quality crystals and gives rise to stray crystals. For these reasons, it is necessary to determine the saturation temperature of the solution T_S (cf. §25) before placing a seed in it, although such a determination is time-consuming and sometimes complicated.

When the value of T_S has been determined, a temperature 5-6 deg C higher than the saturation temperature is established in the thermostat. The crystallizer is sealed hermetically and rested for 10-12 h to dissolve the particles captured from air or those split off from a test seed (such particles can act as crystallization centers). A thermometer must be fitted into the cover. After this period, the temperature is reduced. A crystal holder and a seed are prepared. The crystal holder and the cover are washed in hot water and the seed is rinsed in warm water. After this treatment, the crystal holder and the seed must be placed immediately in the solution.

When the temperature in the crystallizer falls to 2-2.5 deg C above the saturation temperature, the seed is introduced into the crystallizer. This operation must be carried out very rapidly and is best done by two people. One of them removes the cover with the thermometer and the other introduces the crystal holder and then they both hermetically seal the crystallizer. The temperature in the thermostat is reduced to the growth temperature, i.e., about 1 deg C below T_S. The beginning of the growth of a crystal is indicated by reflections of light from its faces. Next, a motor is switched on to set the growing crystal in motion. If, after several hours, there are no stray crystals at the bottom of the crystallizer or on the crystal holder and if the seed crystal does not transform into an intergrowth or a dendrite, we may assume that the first stage of the experiment has been completed successfully. In general, a slight supersaturation or supercooling is not harmful; it simply slows down the process but does not stop it. However, a strong supersaturation is the source of many harmful effects. Therefore, it is always necessary to start from a slight supersaturation and, if the growth is proceeding satisfactorily, to increase the initial supersaturation to about twice its original value. If there is no deterioration in the growth, the supersaturation can be increased still further. However, when we find the limit above which a crystal grows with systematic defects, it is necessary to use a value of the supersaturation below this limit in order to maintain stable conditions and obtain good-quality crystals.

It is very important to determine the correct rate of cooling during growth. This is a difficult task and is usually approached by the trial and error method. For example, we can start with a supercooling by 1 deg C. We find that the growth is satisfactory. After 12 h we reduce the temperature by 0.4 deg C. Everything is still in order. After the next 12 h, we reduce the temperature by a further 0.4 deg C. Now, we observe some inclusions. It follows that in the next run the temperature should be reduced in the third stage only by 0.2-0.3 deg C. Since the appearance of inclusions is affected by the properties of a given seed (two seed crystals grown under identical conditions may differ quite considerably), it becomes clear that the trial and error method of searching for optimum growth conditions may take months. For a crystal with a simple geometrical shape, we can use a ruler to measure externally the approximate dimensions of the crystal and estimate its weight. We can then calculate roughly the actual saturation temperature under given growth conditions. Using such a calculation, we can deduce how far we can reduce the temperature in the crystallizer [Ansheles, Tatarskii, and Shternberg (1945)]. However, it is best to determine the concentration of a solution during a run. This can be done either by measuring the refractive index of the solution using immersible refractometers [Ioffe (1960)], or by measuring the electrical resistivity of the solution (cf. §26).

If this method of growing crystals is to be used extensively in a laboratory, it is desirable to construct a device for automatic cooling; a variant of such a device is described, for example, by Šip and Vaniček (1962). Such devices do not have feedback circuits and therefore they can be used only when no stray crystals are formed; they do not give optimum and reproducible conditions in the presence of stray crystals. Cooling devices which would allow for a deviation of the supersaturation from a set value would be more useful. The authors are not aware of the existence of such devices.

The crystallization process may be regarded as complete after 24 hours from the last reduction of temperature. Then, a crystal is extracted from the crystallizer. The remaining solution can be used again but an amount of the original substance equal to the weight of a newly grown crystal must be added to it. During long runs, the solvent losses may be considerable. Therefore, after filling the crystallizer it is necessary to measure the level of the solution and, at the end of a run and after the addition of a new portion of the substance, solvent must be added up to the previous level of the solution. The saturation temperature is determined again and the process is repeated as described above.

Even when very pure substances are used, the repeated growth of crystals from the same solution results in an accumulation of impurities in the solution which, sooner or later, begin to affect the growth and quality of the crystals. It follows that it is necessary to change the solution from time to time.

In this method of growing crystals, the greatest difficulty is the prevention of the formation of stray crystals.

We shall now consider the main causes of the formation of stray crystals and some methods for preventing this. Single or multiple stray crystals may appear quite quickly (during the first few hours after the beginning of growth of the main crystal) or they may appear in the later stages. In the former case, stray crystals are due to errors in the selection of the correct experimental conditions while in the latter case they are due to a disturbance in the normal growth process. The main causes of the formation of stray crystals due to incorrect initial conditions are as follows:

1) The capture of crystalline dust particles from air during the introduction of a seed into the crystallizer (measures for preventing this are described in §14).

2) Insufficiently careful washing of the cover, crystal holder, and seed crystal.

3) Defects in the seed crystal (cracks and occlusions).

4) Insufficient duration of the storage or too low a temperature of the solution after the determination of the saturation temperature by means of a test seed.

5) Too high a value of the initial supersaturation, which may be due to the evaporation of the solvent (as a result of the poor sealing of the crystallizer during storage before the introduction of the seed) or due to errors in the determination of the saturation temperature.

All these faults can be remedied quite easily but if a crystal film or floating crystals appear on the surface of the solution in spite of the measures that have been taken to ensure the correct initial conditions, it follows that the temperature at the beginning of the process should be altered or the apparatus should be modified. If the hermetic seal is satisfactory, the appearance of floating crystals on the surface of the solution is the result of strong supercooling of the surface. This can be prevented either by additional thermal insulation of the crystallizer cover or by special heating of the cover.

Stray crystals may appear several days after the beginning of the growth of a crystal for the following reasons:

1) The appearance of single crystals (particularly at the same points in the crystallizer in several consecutive runs) indicates cracks or scratches in which residues of the crystal phase may remain even after careful washing (this happens in crystallizers in which a solution is permitted to dry out: such crystallizers should be filled with a solvent and left in the filled state for several days before using in the next run).

2) The formation of many stray crystals usually indicates that the crystallizer is not hermetically sealed.

3) A disturbance in the normal operation of the drive from the motor, which may shake the crystal holder and fracture a growing crystal attached to it.

4) Cracks may appear in a crystal and these may emerge on the surface (cf. §21).

5) The rate of cooling may be too high, resulting in the transition of the solution to the labile state (this is almost always preceded by imperfections in a growing crystal, the appearance of inclusions, the skeletal growth of a crystal, etc.).

Naturally, several of these factors may be acting simultaneously and measures to prevent the effect of one of them may not give any perceptible improvement. When stray crystals are already present we can heat the solution to such an extent as to ensure a higher temperature in the lower part of the crystallizer. This can be done in a liquid thermostat by immersing the crystallizer to a depth of 10-15 mm. In this case, the temperature in the crystallizer will be somewhat lower than the temperature of the thermostat, as in the method in which a temperature difference is used as the driving force of the crystallization process.

The cooling method is fairly complex. However, it is widely used to prepare crystals from solutions, particularly from molten salts. It is used because of the relative simplicity of the apparatus and because this method is known more widely than other techniques. It must be stressed that the nonisothermal conditions of crystal growth in this method do not hold out much promise of its further development.

§13. Crystallization by Solvent Evaporation

In this method, an excess of a given solute is established by utilizing the difference between the rates of evaporation of the solvent and the solute. In fact, the component parts of a solution are sorted out in accordance with the degree of their mutual binding. In contrast to the cooling method, in which the total mass of the system remains constant, in the solvent evaporation method, the solution loses particles which are weakly bound to other components and, therefore, the volume of the solution decreases. In almost all cases, the vapor pressure of the solvent above the solution is higher than the vapor pressure of the solute and, therefore, the solvent evaporates more rapidly and the solution becomes supersaturated. Usually, it is sufficient to allow the vapor formed above the solution to escape freely into the atmosphere. If this results in a considerable loss of the solute or if the solvent is toxic, it is necessary to confine, in some way, the solvent vapor to a closed volume, for example, in a desiccator. This is the oldest method of crystal growth and technically it is very simple. However, this apparent simplicity does not mean that there are no difficulties in the preparation of large uniform crystals.

When crystals are grown by this method, all the available solute is in the solution. As the amount of solvent decreases and the solute precipitates, the concentration of all those impurities whose segregation coefficient in this system is less than unity increases. When the amount of the solvent is reduced to half its original value, the concentration of such impurities in the solution is practically doubled and, if the segregation coefficient remains constant, the concentration of these impurities in a growing crystal also doubles. Therefore, in principle, the preparation of uniform crystals by this method is even more difficult than by the cooling method and, therefore, it is necessary to use specially purified solutes and solvents. Alternately, one can prepare small crystals from a large volume of a solution so that during a given run the volume of the solution does not change greatly and therefore the concentration of impurities in the solution remains practically constant.

Another important factor which affects the uniformity of crystals is the supersaturation. The supersaturation also changes during crystallization. The change in the supersaturation depends primarily on the rate of evaporation of the solvent. The rate of evaporation of a unit mass of the solvent depends, other conditions being equal, on the surface S of the liquid in a crystallizer. The resultant excess of the solute is distributed in the solution whose volume is v. The rate of increase of the supersaturation in the absence of a seed is proportional to S/v. If we use a cylindrical container, the supersaturation changes little if the liquid column in the cylinder is high but it changes very rapidly when the amount of the liquid is small. The dependence of the rate of loss of the solvent on the height of the liquid column is elementary:

$$\frac{S}{v} = \frac{\pi r^2}{\pi r^2 h} = \frac{1}{h},$$

where h is the height of the liquid in the cylinder.

Thus, evaporation of the same volume of the solvent at the beginning and end of a run produces very different changes in the supersaturation: the change is greater when the amount of liquid is less. Therefore, it is best to carry out a crystallization run in such a way that the change of the level of the solution is small compared with the height of the liquid column. The change in the supersaturation depends not only on S/v but also on the surface of a crystal (or crystals) present in the solution. Since the surface of a freely growing crystal increases with time, such a crystal captures more and more of the solute. Therefore, an increase in the supersaturation will be compensated by an increase of the area of the growing crystal. The large area of a crystal (particularly in the presence of several seeds distributed uniformly at the bottom of the crystallizer or elsewhere) also ensures a slower change in the supersaturation and, consequently, a higher quality of crystal. Naturally, an increase in the size of several seeds will be smaller compared with that which would have been obtained in the presence of a single seed. The supersaturation depends also on the solubility of the solute. The higher this solubility, the higher is the supersaturation in the solution at a given evaporation rate of the solvent.

Apparatus used in growing crystals by the solvent evaporation method may take many forms, but there are certain important common features.

1. A crystallizer should be placed in a thermostat. The temperature should be maintained as accurately as in the crystallization by convection methods. However, when the solubility of a solute depends weakly on temperature and high-quality crystals are not required, the temperature need not be maintained very carefully. In such cases, passive thermostating is used, which simply ensures the absence of large fluctuations of temperature.

In slow fluctuations of temperature, an increase of the temperature reduces the supersaturation. However, a temperature rise increases the rate of evaporation and this slows down the decrease of the supersaturation. During cooling, the effects are opposite. Moreover, slow changes in the supersaturation are, generally speaking, less harmful than sudden changes in the crystallization conditions. For example, sudden changes may give rise to inclusions of the solution in a crystal.

2. A solution must be evaporated under "sterile" conditions, i.e., the solution must be isolated from a dusty atmosphere. If this precaution is not taken, many stray crystals are precipitated, which interfere with the growth of the main crystal. The rate of growth of the main crystal slows down and, moreover, if any stray crystals are in contact with the main one, cracking may take place, stresses may be generated at the points of contact, and solution may be included in the main crystal. The apparatus used in the crystallization by the evaporation of a solvent can be seen in the diagrams in the present section. The simplest variant is shown in Fig. 31. Here, the temperature is kept constant by the thermal inertia of the water in the outer container (thermostat).

When a lower supersaturation is required, the number of sheets of filter paper covering the crystallizer (Fig. 31) is increased or the area of the aperture in the cover may be reduced. Moreover, a tall crystallizer, for which the ratio S/v is less, can be used. Finally, several seeds can be placed in the crystallizer. A desiccator (Fig. 32) should be used when the water vapor pressure above the solution is less than the vapor pressure in the atmosphere so that crystals dissolve in air (this happens in the case of NaI); under these conditions large crystals cannot be obtained by the simple evaporation of a solvent directly into the atmosphere.

The rate of evaporation can be controlled as in the preceding case (cf. Fig. 31).

The desiccator technique is as follows.

A solution saturated at the selected growth temperature is prepared by mixing, using published data if necessary. This solution is heated to a temperature 8-10 deg C higher than

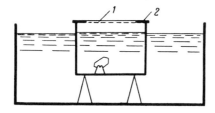

Fig. 31. Crystallizer in a water bath (passive thermostating). 1) Filter paper; 2) annular cover holding filter paper on the top of the crystallizer.

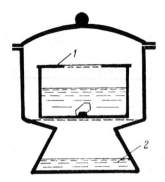

Fig. 32. Crystallizer in a desiccator. 1) Cover which limits the rate of evaporation; 2) sulfuric acid or calcium chloride for absorption of water vapor.

the saturation temperature, and it is poured into a clean (preferable wet) crystallizer. If the solubility of the solute depends strongly on temperature, a crystal holder is placed in the solution when the temperature is 2-3 deg C higher than the growth temperature. It is also possible to pour a solution into a crystallizer containing seeds. Such seeds are best placed in cavities made in a plate of some inert material (polyfluoroethylene or polymethylmethacrylate). The crystallizer is closed by a specially prepared cover and placed in a thermostat or a desiccator. After some initial dissolution, a crystal may appear in a cavity on the second or third day; therefore, patience is required before drawing any conclusions about the success of a run. After having been cooled, the crystallizer should not be moved, inclined, jerked, etc., since the liquid, which would then wet the walls, could give rise to stray crystals.

If the evaporation is accelerated too much, stray crystals may be formed. This is because, under the static conditions which must be maintained in this method, the concentration of the solution vaires considerably from point to point. A strong supersaturation may be found at the surface of the solution and, consequently, the spontaneous formation of stray crystals may take place at the surface. The likelihood of the formation of such crystals is higher for larger volumes of a solution and larger distances between the surface and a growing crystal.

One of the common difficulties of this method is the creep of the solution along the walls of the crystallizer. A solution rises up the walls by capillary forces and crystallizes there to form stray seeds. To avoid such creep, it is usual to smear the walls of the crystallizer above the solution level with some inert substance such as paraffin or some other wax. It is also desirable to reduce the rate of evaporation. The creep of a solution up the walls depends strongly on the acidity of the solution (this has been found by the present authors in the crystallization of KNO_3). Therefore, we can suppress such creep by altering somewhat the acidity of the medium, provided this does not affect the quality of the growing crystals or produce reactions with the solution.

A convenient method for growing crystals in molds by the evaporation of a solvent has been described by Shternberg (1961). The apparatus described by Parvov (1964) relies on forced convection and permits continuous control of the rate of evaporation of the solvent. A crystallization method with a controlled rate of evaporation of the solvent, which can be used for substances with a retrograde temperature dependence of the solubility (such as $Li_2SO_4 \cdot 2H_2O$), has been described by Nepomnyashchaya, Shternberg, and Gavrilova (1962). These authors have also described a dynamic variant of their method. The addition of a solution to a crystallizer changes the method described by Nepomnyashchaya et al. to a steady-state method. A very convenient method, suggested by Karpenko et al. (1961) and improved by Sampson and di Pietro (1963), is based on the evaporation in a closed container using the temperature drop between the liquid and the vapor above the solution. This method is also of the steady-state type.

Simple variants of the solvent evaporation method are very convenient when small crystals are required in investigations of various physical properties. Crystallization of small amounts of the solution from watch-glasses or from Petri dishes can be used to rapidly obtain preliminary data on the effect of the chemical nature of a medium on the growth of crystals (cf. §20).

CHAPTER III

TECHNICAL EQUIPMENT FOR A CRYSTAL-GROWING LABORATORY

§14. Laboratory Buildings and General Equipment

A laboratory in which crystals are grown should satisfy the same requirements as an ordinary chemical laboratory. It needs: an exhaust (fume) hood; running water; a floor of washable plastic tiles free of gaps at the joints. The laboratory should have a 24-hour failure-proof electrical power supply because the process of growing crystals may take weeks or months. The benches on which the crystallizers are placed should have a fireproof covering and each unit should have its own fuse in case of a short in the heater circuit. Work on dry substances should be carried out under an exhaust hood or, better still, in a separate room.

The laboratory should be equipped with the following apparatus:

1) A distillation unit of the D-1 type.

2) Desiccators for drying dishes and reagents. It is also highly desirable to have a vacuum drying chamber (VSh-0.035) because reagents which cannot be heated can be dried rapidly in this chamber. Such a chamber should not be used for the evaporation of solutions because the vapor may condense in the pump.

3) General-purpose and analytic balances.

4) A gasoline burner with a foot bellows, or better still, with a compressor, for soldering, filling, and other laboratory work. Such a burner is used to prepare crystal holders, ampoules, test tubes, and other simple glassware. A bellows or a compressor can also be used to accelerate filtration (see below).

5) A universal electrical meter or tester, type TT-1, Ts-20, AVO-5, or some other. Such a meter is used to measure the various parameters of the electrical circuits and in assembly of relay circuits and heaters.

6) A pH meter, type LP-58 or LPU-01, for measuring the acidity of solutions.

7) A binocular microscope, type MBS-1 or MPS-1.

8) A chamber for work with strongly reactive corrosive and radioactive substances. It can be used also for work with toxic substances and for grinding and weighing dry substances before the preparation of solutions.

Apart from such general-purpose equipment, a crystal-growing laboratory should possess apparatus for the preparation of uniform single crystals. These requirements apply also when the main work of the laboratory is concerned with studies of defects in crystals. These special requirements are:

1) It should be possible to carry out crystallization using a driving force (external agent) of a high degree of constancy;

2) A crystallization medium should contain the minimum possible number of foreign substances in the dissolved and suspended state, with the exception of those substances which are introduced deliberately into a solution.

These requirements can be satisfied using the apparatus described in the following sections of the present chapter.

§15. Thermostats

We have already demonstrated how the temperature affects the driving force of crystallization $\Delta\mu$. Therefore, it is clear that a highly stable temperature is one of the basic prerequisities for the successful growth of crystals.

When the crystallization is carried out at room temperature, the simplest method for increasing the stability of the temperature in a crystallizer is to surround it by a thermally insulating jacket or to place it in a bath containing a large amount of water (10-20 liters); however, in such a thermostat the temperature fluctuations are not eliminated but simply reduced compared with those in the laboratory. Therefore, such thermostats should be placed as far as possible from furnaces, ovens, windows, and doors. These thermostats are best placed in a basement or in an unheated windowless room. This thermostating method (known as the passive method) can be used to grow crystals by chemical reaction under counter-diffusion conditions and to grow small crystals by other methods when good quality is not of great importance.

When the requirement is for large uniform crystals, active thermostating is necessary, i.e., we must use devices which control the temperature (Fig. 33). The general design of a thermostat with a control system includes a thermally insulated chamber 1, a pickup 2 which responds to changes in temperature and feeds back an electrical signal to a control circuit (for example, it might send a signal that the temperature is higher than the set value). The control circuit 3, which receives this signal, transforms it, amplifies it, and takes appropriate action (in this case, switches off a heater 4).

Let us now consider each part of a thermostat in turn.

1. Figure 34 shows a thermostat which is employed in crystallization by cooling or by thermal convection. It is in the form of a thick-walled cylindrical glass bath filled with water (at temperatures above 70°C oil is used). The thermostat is covered. A rubber seal is placed between the cover and the bath. The bath is placed on a base and the cover is screwed tightly against the base. The thermostat cover has an aperture for a crystallizer, which is supported on a separate base 9. The rods which support this base pass through the thermostat cover. The position of the base 9 can be varied by screws 3. The crystallizer is sealed hermetically by a cover 4 fastened to its top by screws 2. The screws and nuts used to hold the crystallizer base should be made of brass. A combination of brass and iron parts immersed in water is not permissible even if these metals are not in contact because this combination results in the oxidation of the iron and the contamination of the bath water.

The volume of the thermostat is governed by the dimensions of the crystallizer and the accuracy with which the temperature must be kept constant. The larger the volume of the thermostat, the smaller are the fluctuations of the temperature

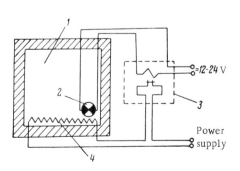

Fig. 33. General layout of a thermostat with a control device.

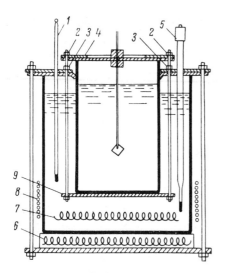

Fig. 34. Thermostat used to grow crystals by cooling a solution.

in the crystallizer, other conditions being constant. In the thermal convection method, the thermostat need be only 1.0-1.5 liters in volume but in the case of crystallization by cooling the thermostat volume should be 10 liters or more. Apart from the crystallizer, the thermostat cover also carries a thermometer 1 and a heat controller 5.

To increase the accuracy of the maintenance of the temperature at a constant value and to improve the uniformity of the temperature field, the water in the thermostat must be stirred. However, stirring has its own disadvantages: the construction of the thermostat is more complicated, there are additional possibilities of breakdown (failure) and noise. For these reasons, when it is necessary to maintain the temperature accurately at some set value, it is best to use a "double" thermostat in which the thermostat bath and the crystallizer are separated by another bath filled with water. In other words, we introduce an intermediate water jacket. The efficiency of a double thermostat is not inferior to stirred thermostats. The walls of both thermostats should allow an investigator to observe visually the crystallization process and, therefore, the crystallizer cannot be placed into an industrial Wobser-type TS ultrathermostat or a similar unit. The use of ultrathermostats is permissible only when each is used to maintain a temperature in several glass baths in which crystallizers are placed. The thermostats used in crystal-growing laboratories are usually made in their own workshops.

Industrial-type air thermostats and desiccators can be used in a crystal-growing laboratory when the temperature need not be maintained very exactly. In preliminary experiments on crystals growing by solvent evaporation, by chemical reactions, etc., it is frequently necessary to construct special thermostats which are more suitable for a given experiment than standard thermostats. They are usually made of asbestos cement, plastics, glass, or plywood. Temperature fluctuations in such thermostats can be reduced, as in water baths, by stirring the air in them or by using double thermostats. The walls of the inner thermostats can be made of Plexiglas. Air thermostats must not be opened when they contain crystallizers. Therefore, such thermostats should also have viewing windows and lamps for illuminating the interior.

When a large number of crystallizers is used under the same conditions, it is desirable to construct air thermostats of large dimensions in the form of special chambers or even whole rooms.

2. Contact thermometers are usually employed to measure the temperature. To maintain a constant temperature throughout the whole experiment, one can use contact thermometers with fixed contacts. Such thermometers are generally available for various ranges of temperatures. The most convenient contact thermometers are of the TK-6 type, with which it is possible to vary the temperature continuously.

All contact thermometers are based on the same principle. Electrical contacts are placed at different heights in a capillary. When the temperature rises above the set value, fixed by the position of the lower contact, the rising of the mercury present in the capillary closes the two contacts.

Instead of a contact thermometer we can use an electronic temperature controller of the ÉRA-1 type, in which a resistance thermometer is the sensing element. In air thermostats, in which the temperature needs to be maintained to within ±0.5 deg C, we can use bimetal strips as sensing elements. A list of commercially available sensing elements is given in the Automatic Devices and Controllers handbook (1964).

3. The simplest variant of a telephone relay can also be used as a control unit; such a relay has contacts which are closed when a coil surrounding them is open-circuited. As long as the contacts in a contact thermometer remain open, there is no current in the relay coil and the closed contacts of the relay permit the passage of a current through the heater. When the temperature rises to a level sufficient to close the contacts in the thermometer, a current appears in the relay coil which opens the contacts in the heater circuit. The temperature is controlled in the same way when bimetal strips are used as the sensing elements.

The long periods necessary to grow crystals and the need to avoid any interruption of the temperature control process means that special measures must be taken to ensure that the control system is reliable. In the control system, there are two aspects which require special attention. The first is the contact between the mercury and the metal wire in a thermometer. When a spark jumps at the moment of breaking the circuit, the temperature of the mercury meniscus rises and the mercury evaporates. Moreover, after many electrical discharges of this type the walls of the capillary become covered with a layer of mercury. The form of the meniscus changes and, therefore, sooner or later, the moments of making and breaking of the contact cease to correspond exactly to the set temperature. Therefore, the most radical means of increasing the reliability of the operation and the long service of such sensing elements is to ensure the minimum possible value of the current flowing through them. The maximum electrical power which is stated as the permissible limit in the factory description of the TK-6 is 0.2 W, which is far too high for the long service life of a temperature controller.

Another weak point of the control system are the contacts in the relay which switches on the heater. The use of a relay with contacts capable of carrying only low powers results in the rapid burning out of the contacts and a resultant lack of reproducibility of the switching of the heater. Here, we must have a three- to five-fold margin in the permissible power compared with the power of the heater.

However, if measures are taken to improve the reliability of these weak points, we find that the requirements are contradictory. The point is that the weaker the signal produced by the sensing element, the lower is the power that reaches the relay. The lower the power reaching the relay, the weaker is the action of the contacts. Therefore, we have to use either an intermediate relay which amplifies the primary signal or we must use vacuum-tube or semiconductor circuits for amplification of the sensing-element signal. A very reliable circuit has recently been developed for this purpose; it is shown in Fig. 35. The control (auxiliary) relay 1 can be any low-power relay operating satisfactorily under a constant voltage of 12-24 V. The terminals of relay 1 close the circuit of the coil of the power relay 2. The terminals of relay 2 should be capable of passing a current of 2-3 A without pitting. Resistors R_1 and R_2 ($R_1 \ll R_2$) are selected in such a way that R_1 has a minimum value for the reliable operation of relay 1, while R_2 has a maximum value. We can use RPN and RKS relays for the relays 1 and 2, respectively. In this case, $R_1 \approx 100$-200 Ω and R_2 can be up to 10 kΩ. Such a circuit is supplied with power from a rectifier of the VS-24 type.

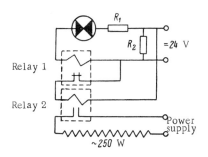

Fig. 35. Circuit of a simple temperature-controlling system with two relays.

The factory producing the TK-6 contact thermometers recommends the semiconductor circuit in Fig. 36.

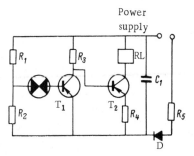

Fig. 36. Circuit of a thermal controller in which transistors are used. T_1 is a P4 transistor; T_2 is a P13 transistor; D is a DGTs-27 germanium diode; C_1 is a KÉ-1 electrolytic capacitor (20 μF; 50 V); RL is an MKU-48-S electromagnetic relay (24 V); R_1 is a VS-0.25 resistor (110 kΩ); R_2 is a VS-0.25 resistor (2.2 kΩ); R_3 is a VS-0.25 resistor (5.5 kΩ); R_4 is a VS-0.5 resistor (47 Ω); R_5 is a voltage-dropping resistor.

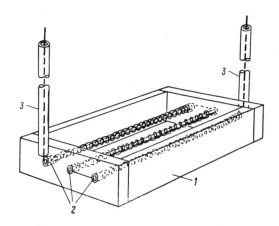

Fig. 37. Heater located within a thermostat. 1) Plastic frame; 2) glass rods on which a helix is wound; 3) heater leads in glass tubes.

4. When liquid thermostats are used, the heater is usually in the form of an Nichrome helix, which may be located in a number of places in a thermostat.

a) A heater may be placed under the base of a thermostat bath in a special asbestos jacket built in the same way as electric room heaters (6 in Fig. 34). This is the most reliable and convenient method of heating. However, in this case, the temperature in the thermostat fluctuates within relatively large limits: in large thermostats (in the case of crystallization by cooling), the fluctuations may be up to 0.3 deg, while in small thermostats (used to grow crystals by thermal convection), the fluctuations may reach 0.5 deg C. The temperature fluctuations in crystallizers are much smaller.

b) A helix may be wound externally around a beaker (8 in Fig. 34), and attached to it with a silicate adhesive. For safety reasons, it should be covered by a layer of asbestos. Temperatures fluctuations are approximately the same as in the preceding case but such a heater is easier to make.

To avoid cracking of the thermostat beaker by thermal shock, it is necessary to avoid heating currents which would make the heater wire glow red in cases "a" and "b". In case "a", it is also desirable to leave a gap between the heater and the bottom of the bath.

c) A heater may be placed directly in water at the bottom of a beaker (7 in Fig. 34). A heater helix should be uniformly distributed along the bottom of the thermostat but should not touch it. This is the most economical way of heating. Temperature fluctuations are minimal but the heater wire is relatively rapidly corroded by water. The corrosion is accelerated considerably when a solution of the substance being crystallized is accidentally dropped into the liquid used in the thermostat. The ends of the heater wire should project through opposite sides of the thermostat. This reduces the current passing through the water and helps to prolong the life of the heater. The contacts of the heater with the supply leads should be outside the thermostat (in air). A variant of such a heater is shown in Fig. 37. A thermostat is best filled with distilled water. In the case of liquid-filled thermostats with low walls (when crystals are grown by the thermal convection method), incandescent lamps can be used as heaters. One must make sure that the liquid does not come into contact with such lamps.

The power consumption of a heater (its resistance) is selected so that the duration of the heating is approximately equal to the duration of cooling of the thermostat. Naturally, at different temperatures, the power should be different and, therefore, it is most convenient to provide each heater with its own autotransformer or a rheostat. The power of a heater for a large thermostat of 10-15 liter capacity should be 150-300 W.

The heating elements in air thermostats are usually Nichrome helices or incandescent lamps (the latter are safer), placed along the perimeter of the lower part of the thermostated volume.

A very important factor are the relative positions of the temperature-sensing element and the heater. The closer these two units are, the more accurate is the maintenance of a constant temperature, but the frequency of operation of the contacts in the sensing element and their wear also increase. When the sensing element is far from the heater, the frequency of operation of the element decreases but the stability of temperature maintenance is lower. Thus, the quality of the thermostat is only partly governed by the sensitivity of the sensing element.

§16. Devices for Producing Relative Motion in a Crystal – Solution System

The forced motion of a solution past a crystal makes it possible to increase considerably the rate of its growth without deterioration of its optical homogeneity. There are many ways of setting a solution in motion. The transmission of any motion within a crystallizer presents problems because of its hermetic sealing. If a crystallizer is not hermetically sealed, uncontrolled changes in the concentration of a solution may take place, stray crystals may be formed, etc.

All the methods of setting a crystal in motion can be divided into two groups, in accordance with the hermetic sealing method.

The first group includes various forms of motion transmitted to a stirrer or a crystal through various seals. Such seals must be as tight as possible and they must prevent contamination of the solution with the lubricating or thermostat oil or the products of friction of various parts.

Liquid seals are not used. The reason for this is that the mercury used in such seals is corrosive, and it is difficult to find other liquids which would not creep or react with solutions and which would not be volatile. For this reason, the various seals employed currently have ball-bearings with packing glands. In the case of corrosive liquids, the packing glands are made of Teflon in combination with Plexiglas. Packing glands made only of Teflon cannot be used because, in the case of the friction of Teflon against Teflon, flakes break away and the hermetic seal is quite rapidly lost. It is best to construct a separate seal for each application. The reader is advised to refer to a book by Chernousov, Kutin, and Fedorov (1965) for methods of hermetic sealing.

We shall now consider various forms of mixing or stirring the solution, in which such seals are used.

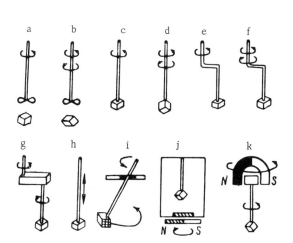

Fig. 38. Various ways of producing relative motion between a crystal and a solution.

1. A solution can be stirred by rotating, in one direction, a paddle located

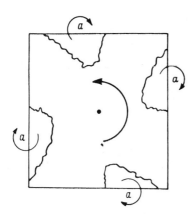

Fig. 39. Distribution of inclusions of the mother liquor in a crystal rotated in one direction. Steady-state eddies in the solution are denoted by "a."

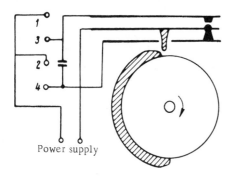

Fig. 40. Device used for the periodic change of rotation of the axle of a motor. The numbers refer to the numbers of the terminals of ASM-type motors.

centrally above a crystal fixed to the bottom of a crystallizer (Fig. 38a).

2. We can employ a centrally located stirrer whose direction of rotation is periodically reversed (Fig. 38b).

3. A crystal attached to a holder can be rotated in one direction about the holder axis (Fig. 38c). This method is employed relatively rarely because inclusions usually form in the crystal (Fig. 39), and these inclusions are located immediately behind the projecting edges of the crystal (cf. §6).

4. A crystal attached to a holder can be rotated clockwise and anticlockwise about the holder axis (Fig. 38d). In this method, the direction of rotation of the crystal is changed periodically (usually every 10-20 sec) and inclusions are not normally formed. A periodic change of the sense of rotation of the axle of a motor is possible if a three-phase drive (for example, of the ASM type) is used with a special device shown in Fig. 40. A cam changes the phases of the power supplied to a quiet-running low-power motor (SD-2 or DSM-2) rotating at 2 rev/min.

5. One can also employ eccentric unidirectional motion of a crystal (Fig. 38e) [Ansheles, Tatarskii, and Shternberg (1945)].

6. Eccentric clockwise and anticlockwise motion of a crystal can also be employed (Fig. 38f).

7. Planetary rotation of a crystal is sometimes used (Fig. 38g). This method has been developed by Mokievskii (1948). It differs from other methods in that the crystal is set in motion by means of a special mechanical device along a circle in a solution and, simultaneously, the crystal rotates about its own axis. We can see no special advantage in this method over the clockwise and anticlockwise rotation. Mokievskii's method is more complex and a mechanical device placed in the crystallizer above the solution may be a source of contamination.

The rates of motion used in central rotation are usually 60-200 rev/min. The rates of motion in eccentric and planetary rotation usually do not exceed 100 rev/min. Higher rates of motion, particularly in the case of large crystals, may be harmful. Crystals may crack next to the crystal holder. Moreover, the likelihood of the formation of stray crystals is higher when the stirring rate is increased. This may be due to the cracking or scratching of the surface of a crystal by particles present in a not very clean solution, or it may be due to cavitation which tends to produce sharp edges in a crystal. Furthermore, an increase of the rate of motion can produce only a limited increase in the rate of growth (§6).

The second group of methods of producing relative motion between a solution and a crystal comprises various techniques in which no seals are used. These methods make it possible

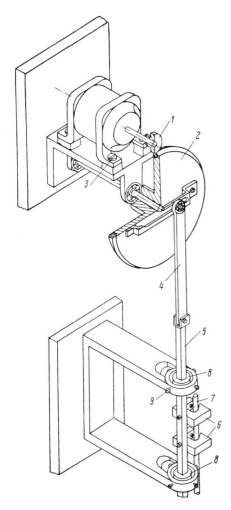

Fig. 41. Device used to set a crystal-holder in reciprocating motion.

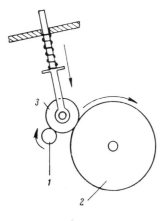

Fig. 42. Device with an idler pulley: 1) driving pulley; 2) driven pulley; 3) idler pulley; 4) spring.

to seal a crystallizer completely and to attain the minimum contamination of the solution.

8. A reciprocation or vibrational motion (Fig. 38h), described by Petrov (1960) and by Šip and Vaniček (1962) is as effective as any other method. The best frequencies are 2-5 cps and the best amplitudes are 10-25 mm. The device, shown schematically in Fig. 41, works on these lines. The frequency of vibrations is governed by the rate of rotation of the axle of a motor and the ratio between the radii of the driving (1) and driven (2) pulleys. Using different driving pulleys but the same motor and the same driven pulley, the frequency can be varied within fairly wide limits. The tightness of the contact between the pulleys is controlled by screws 3. The amplitude is varied by moving the upper end of a crankshaft 4 along a groove in a plate fixed to the driven pulley. A rod 5, to which the clamps 6 of a crystal holder 7 are attached, moves in rings 8 which are centered by the screws 9.

To grow crystals up to 1 kg in weight, we can use motors providing 5-10 W to the axle. It is best to use brushless motors with ball-bearings since these motors have a longer service life, their rate of rotation is more constant, and they are less noisy. ASM-type motors (already mentioned) are convenient for this purpose. The friction coupling between the pulleys is best achieved by using the following pairs of materials: rubber and glass or plastic and metal. A pulley rotating at a higher rate is best made of a low-wear material. A more reliable method than direct friction between the axle of the motor and a pulley is a system with a rubber idler pulley which, as it wears down, is pressed by a spring to the axle of the motor and to the driven pulley (Fig. 42). This system ensures that the frequency of vibrations is constant in spite of wear in the idler pulley.

Gear or worm speed reducers are more reliable but more complex. To achieve quiet working of such gear reducers, it is best to use plastic pinions with helical teeth.

The length of a crankshaft in the device shown in Fig. 41 should not be less than 100-120 mm. Otherwise, the wear of the centering rings is rapid and the uniformity of motion deteriorates. The rate of motion of a crystal relative to a solution varies almost sinusoidally, with a maximum near a point

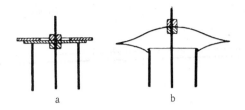

Fig. 43. Two types of hermetic cover for crystallizers.

equidistance from the point where the motion is reversed. The maximum velocity of a crystal is approximately equal to the linear velocity along a circle of radius r whose center coincides with the bearing of the crankshaft 4 attached to the pulley 2 (Fig. 41). This velocity (v_m), which is of decisive importance in the formation of a diffusion layer around a crystal, is found from the formula

$$v_m = \frac{2\pi r}{\tau_0} = \pi A \nu,$$

where τ_0 is the period of revolution; A is the amplitude of a vibration; ν is the frequency. The longer the crankshaft compared with the amplitude of vibrations, the more accurate is the value of v_m given by this formula. It follows from this formula that a particular velocity may be obtained by any suitable combination of the values of A and ν. Therefore, we can attach a cam to the axle of a motor without a reducer and obtain frequencies of 20-40 cps by a corresponding reduction of the amplitude so that $v_m \approx 20$ cm/sec. However, under such conditions, large stresses, depending on the mass of the crystal, are applied to the crystal when it changes its direction of motion. Such frequencies are used but only when small crystals, of up to a few grams in weight, are being grown.

The hermetic sealing of a crystallizer in this method of stirring is relatively simple. If the crystallizer is of relatively large diameter ($\gtrsim 10$ cm), the cover is in the form of a rubber membrane (Fig. 43a) in which the crystal-holder is rigidly fixed by means of rubber stoppers. In the case of smaller crystallizers, double membrane configurations (Fig. 43b) or bellows are used. Such double membrane units are made by sticking together the edges of two circular pieces of thin rubber in which an aperture is made for stretching the unit onto the edge of the crystallizer.

The same group of methods of producing motion includes a technique suggested recently by Voitsekhovskii (1963); it is shown in Fig. 38i. Here, a hook, which pulls a crystal holder, is attached to the axle of a vertically fixed quiet-running motor. Since the crystal holder is attached to the elastic cover of the crystallizer (the cover is made of rubber or polyethylene), the crystal describes circles in the solution and the motion is of the planetary type. This method can be used only for crystals grown in a beaker whose dimensions are much larger than the dimensions of the crystal.

Other ways of generating motion in which a crystallizer is sealed hermetically include a magnetic stirrer (Fig. 38j) and a magnetic method for rotating the crystal holder (Fig. 38k). However, recent work has shown that the physical properties of aqueous solutions change considerably in magnetic fields. It is not clear how magnetic fields affect the growth and quality of crystals.

Popov and Sheftal' (1946) have suggested using the inertia of a liquid in a crystallizer without any moving parts. In this method, the whole crystallizer is set in to-and-fro rotation about a vertical axis. The crystal holder then moves at a varying velocity with respect to the liquid. The frequency of the change of direction of rotation is governed by the viscosity of the liquid, the shape and dimensions of the crystallizer, and the velocity of rotation. Because of the complexity of this dependence, and particularly because it is not possible to observe visually the crystallization process or the nature of motion of a liquid at high temperatures in crystallizers made of opaque materials, it is necessary to select empirically the frequency of reversal of the direction of rotation. For this purpose, one has to model the stirring using liquids whose low-temperature viscosity is similar to that of the solution or the melt in which crystallization is to be carried out; this method was rediscovered and used in growing yttrium iron garnet crystals from molten $PbO-PbF_2$ at 1100-1300°C [Titova and Petrov (1965)].

§17. Filtration Devices and Methods

The filtration of solutions is necessary to free them from particles which may scratch and damage mechanically a rapidly moving crystal. The presence of solid particles may also split off small fragments from a growing crystal, which can give rise to stray crystals. Moreover, foreign particles may themselves act as growth nuclei.

The filtration of solutions used in crystallization presents several difficulties. The basic difficulty is that supersaturated solutions cannot be filtered. The filtration of supersaturated solutions usually starts crystallization in the filter and the liquid ceases to pass through it. Crystallization also continues after the solution has passed through a filter. Consequently, a filtered solution must be poured into another container and heated again to dissolve the precipitate. All these operations result in excessive loss of the solvent and solute. Therefore, only unsaturated solutions, heated to a temperature 10-15 deg C above the saturation temperature, are filtered. Heating to temperatures much higher than the saturation temperature is undesirable because it results in considerable losses of the solvent and a change in the concentration of the solution.

Naturally, one should aim to make the filtration process as short as possible. For this reason, it is usual to filter solutions into a flask in which the pressure is below atmospheric. Air is pumped out by a water-jet pump 1 (Fig. 44). A Drechsel washer flask 2 is used to prevent tapwater from reaching the filter flask 3 when the rate of motion of the water drops accidentally due to a drop in pressure in the waterpipe. A stopcock 4 is used to cut off the pumping system when the amount of water in the Drechsel washer flask reaches a dangerous level or when it is necessary to stop pumping at the end of the process. All connections are made by means of thick-walled rubber tubes.

Such a filtration system operates as follows. The filter flask 3 is heated to the temperature of the solution. The filter flask is wrapped in a piece of cloth and connected to the pumping system by a short rubber tube 5. This rubber tube should be fixed permanently to the exit from the flask to prevent the exit glass tube from being broken by an accidental knock. A porcelain funnel 6 is placed into the neck of the filter flask and the funnel passes through a well-fitting rubber stopper. The water-jet pump is then started. The stopcock 4 is still closed. A sheet of filter paper wetted in distilled water is placed on a perforated porcelain plate in the funnel. The edges of the filter paper should not project beyond the edges of the perforated plate. If the sheet is of suitable size and is placed carefully so as to cover all the apertures, the opening of the stopcock 4 should produce the characteristic noise of air being sucked through a wet filter paper. Next, the solution is poured into the funnel. During such filtration, the filtered liquid detaches fibers from the lower side of the filter paper. The number of such detached fibers can be reduced by refiltering the first batch of the filtrate. To do this, after the end of the filtration of the first batch of the solution, the stopcock 4 is closed, the rubber stopper and the funnel are removed, and the filtered liquid is poured back into the solution without

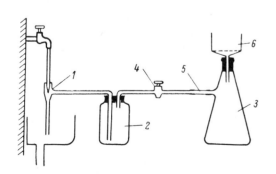

Fig. 44. Apparatus for the filtration of a solution using a water-jet pump.

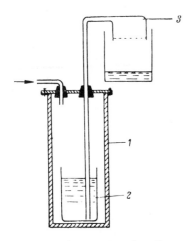

Fig. 45. Apparatus for the filtration of a solution under pressure.

disconnecting the filter flask from the pumping system. The funnel with the same filter is replaced in the stopper and the filtration is repeated. This procedure must be carried out quite rapidly.

If the solution to be filtered is corrosive it is necessary to use filters with porous glass plates. Such filters have numbers stamped on them. The higher the number, the finer are the pores. Consequently, other conditions being constant, the higher the number of such a glass filter, the lower is the filtration rate.

This method of filtration is unsuitable for volatile organic liquids since it results in a considerable loss of the solvent and a change in the concentration of the solution. In this case, one should use filtration under pressure. The apparatus used in such filtration is shown schematically in Fig. 45.

Here, 1 is a metal can which can withstand a pressure up to 3 atm. The liquid to be filtered is placed in a beaker 2. An inverted filter funnel with a glass plate 3 is fixed into the cover of the metal can by a rubber stopper which tapers in the direction of the lower pressure so that it is self-sealing. The cover of the metal can also carries a tube through which air or some other gas is pumped into the can. To establish a pressure inside the can, we can use (§14) a compressor such as that employed for gasoline burners.

In some cases, a heating spiral is wound around the porcelain funnel or the glass filter. Heating is controlled by a rheostat or an autotransformer.

Other filtration methods are described by Lucas (1962) and Berlin (1963).

§18. Materials Used in Crystal-Growing Apparatus

The materials used to make crystallizers must satisfy more stringent requirements than those used in ordinary chemical apparatus. This is because crystals are extremely sensitive to the presence of some substance in the crystallization medium, particularly when the crystallization process is long. The prolonged process means that the products of the slow interaction of solutions with the walls of the crystallizer and with other parts in contact with the solution may accumulate in large amounts in the solution. Therefore, it is necessary to bear in mind the consequences of the interaction of a solution with those parts of the apparatus which will be in contact with it. Let us consider materials used in current practice.

Glass is the most usual and most easily available material for the making of crystallizers and crystal-holders. In most cases, crystallizers are the so-called battery jars. They are available in capacities ranging from half to five liters. If the crystals are to be produced under thermal convection conditions, it is convenient to employ mass-produced tubes of the type used to collect blood from donors. Crystal holders are usually rods or sealed tubes, 5-6 mm in diameter. The thermostats are usually circular baths of 10-20 liter capacity. Thermostat baths can be made of large carboys with their narrow tops cut off. The edges are then ground with carborundum powder on a large sheet of glass.

Chemical glass does not react with organic, neutral, and the majority of acid aqueous solutions. However, solutions of phosphoric and hydrofluoric acids interact with glass. Moreover, glass is not stable when in contact with alkaline solutions. Molybdenum glass is quite rapidly attacked by solutions containing iodine (this is particularly important in the crystallization of KIO_3). It must be stressed that the rate of dissolution of glass increases rapidly when the temperature is increased: the increase is by a factor of 1.5-2.5 for every 10 deg C up to 100°C. Glass is quite unsuitable for temperatures above 150-200°C when high steam pressures are used. At these temperatures glass is not only dissolved, but it also suffers a rapid crystallization, which makes it opaque. The stability of laboratory glass is discussed in a book by Dubrovo (1965).

Before using glass in media which are not very corrosive, it is recommended that the glass is treated for several hours in the hot solution which is to be used for crystallization.

This treatment removes from the surface layer the most easily soluble components and produces a protective silicon layer on the surface. This procedure should be followed with other materials. Such an operation removes the number of impurities which are leached out of the parts in contact with the solution in which crystallization is to take place. For the same reason, it is recommended not to replace, without sufficient cause, the containers and parts which have been used in working with a given solution.

Rubber in the form of sheet is used in hermetic seals for crystallizers; crystal holders which are used to insert seed crystals, etc. are encased in rubber tubes. Rubber is suitable for work with inorganic substances but not for organic materials. Even if there is no external sign of damage to the rubber parts, the consequences of its use in crystallization may be quite serious.

Plastics can be and are used in apparatus employed in growing crystals. The highest chemical stability of all known plastics is exhibited by Teflon (polytetrafluoroethylene). It is employed to make stirrers, crystal holders, seals, etc. The material is not used very widely because it is difficult to assemble parts made from it. It cannot be glued by any of the known adhesives without preliminary treatment of its surface in molten alkaline metals or their solutions in ammonia. Teflon may be pressure-welded but the welding technique is fairly difficult under laboratory conditions. Usually, one has to be satisfied with mechanical fixing of the various parts made from this material.

Plexiglas (polymethylmethacrylate) is frequently used in work with inorganic substances. It is transparent and does not deform up to +60°C; parts made from it are easily glued together by a solution of Plexiglas in dichloroethane; it is easily machined, and is quite stable mechanically. These properties facilitate its wide use in various parts of the apparatus employed in studies of crystallization or in the growing of crystals. The disadvantages of Plexiglas are that when exposed to light it acquires a yellowish hue, and that it is easily scratched because it is soft. Getinax (laminated Bakelite-type material), Textolite (resin-impregnated laminated cloth), and glass-Textolite composite material are not used in solutions because they swell and deform. However, they can be used, in the form of 8-15 mm thick sheets, to make crystallizer and thermostat covers.

Celluloid (dissolved in acetone) and BF-2 glue are used as adhesives in the attachment of small crystals of inorganic substances to crystal holders.

Polyvinyl chloride tubes, 2-4 mm in diameter, are used as seed-crystal holders.

Metals are used relatively rarely in the making of crystallizers and other parts which are in contact with solutions.

Titanium can be used in electrolytic solutions because it reacts with fewer chemical substances than does stainless steel.

Platinum is used in exceptional cases, e.g., in the crystallization of very corrosive solutions when a very high purity is required.

A new material can be used only if a special check is made first on how the products leached from such a material by a solution, or the products resulting from a reaction of a solution with this material, affect the crystallization process. To run this check, it is necessary to compare the crystals prepared from a pure solution with the crystals prepared from a solution kept for a long time in contact with filings or shavings of the new material. If such a test does not produce clear-cut results, quantitative data can be obtained by comparing the rate of growth of crystals from pure solutions and from solutions which have been in contact with a new material. If the rates of growth are similar in both cases and the number of inclusions does not differ greatly, such a new material can be used. General information on materials used in chemical apparatus is given by Lashchinskii and Tolchinskii (1963) and in the book "New Technological Materials" (1964). The relevant literaure is cited in these two books.

CHAPTER IV

PREPARATION FOR AND CONTROL OF GROWTH OF CRYSTALS

§19. Acquisition of Information on a Substance and Selection of a Solvent

Before growing any crystal, we must search the published literature for the method of preparation. First, we should look through an old but still valid five-volume handbook of Groth (1906-1919), which gives methods of growth, describes the external shape and symmetry of crystals and gives the physical properties (mainly optical) of a large number of substances. More recent information can be found by using subject indexes to abstracting journals on chemistry and physics, particularly *Chemical Abstracts*. Information on crystals and methods of preparation, with references to original work, can be found also in the well-known chemical encyclopedias of Mellor (1922-1964) and in the Gmelins Handbuch (1924-1963). Information on organic substances can be found in the reference book compiled by Tal'nikova and Cherkasova (1961).

Extensive information on the growth of crystals is given in the monographs by Kuznetsov (1953), Buckley (1951), Wilke (1963), Jona and Shirane (1962), as well as in the series of Russian collections "Growth of Crystals", translated into English (1957-1968) and in annual "Reviews on Crystallization" published in *Industrial and Engineering Chemistry* (1946-1966). Information on the preparation of crystals is frequently given also in textbooks on crystal structure. Therefore, it is recommended that a search be made in the many-volume *Structure Reports,* which appeared up to the sixth volume under the name of *Strukturbericht;* "Crystal Data" (1963) should also be consulted.

If information is found in any of these sources it can be counted as a success but in the majority of cases the published information is insufficient. Usually, the descriptions of the methods of the preparation of crystals are very brief. If, for some reason, the published method cannot be used, another method must be employed (we shall discuss the selection of the method later).

In such a search, it is necessary to record the information on the external form (habit) of a crystal, its optical and other properties, etc. Such information may be useful in the identification of prepared crystals.

If there is no published information on the growth method it is necessary to obtain information on or determine those properties of the substance to be grown which are important in the selection of the most convenient method and the optimum conditions for growing crystals. The most important properties are as follows.

1. The chemical stability of a substance both in the solid and liquid (dissolved) state should be known: the important aspects are whether a substance has a tendency to resinification, hydrolysis, photolysis, etc. Possible explosive tendencies and flammability should be recorded. Information on the chemical properties and references to original work will be found

in the chemical encyclopedias referred to earlier, in Chemist's Handbook (1962-1966) as well as in numerous textbooks on chemistry, monographs on particular groups of chemical compounds, and in the original literature on chemistry which can be found in the subject indexes of abstracting journals.

2. The toxicity of a substance is important. The toxic properties of many substances are given in "Harmful Substances in Industry" (1963).

3. The solubility of a substance in various media should be known; this is important particularly when the method of preparation from any particular solvent is described.

Information on the solubility can be found, as well as in the chemical reference works already mentioned, in a series of special handbooks: "Handbook on Solubility of Salt Systems" (1953-1963), "Handbook on Solubility" (1961-1963), Seidell's "Solubilities of Organic Compounds" (1941), and Seidell and Linke's "Solubilities of Inorganic and Organic Compounds" (1952) and their "Solubilities of Inorganic and Metal-Organic Compounds" (1958, 1965). Tables on the solubilities of many common substances in water are given in the "Handbook of Chemistry and Physics" (Vol. 2), which is published annually, and in Vol. 3 of "Chemist's Handbook" (1963). Moreover, the last two handbooks give tables of the physical properties of chemical compounds which list solvents applicable to a given substance but these data are frequently given without any quantitative values. If there is no information on low-temperature solvents, it may be useful to consult "Handbook on Fusibility of Systems of Anhydrous Inorganic Salts" (1961). A molten salt may be used as a solvent.

It will be found that the information on the solubility of the same solute-solvent pair will differ from one source of information to another. This is because the determination of the solubility is, in spite of its apparent simplicity, very time-consuming and there are many potential sources of errors. Therefore, the true solubility is usually not known with an accuracy of better than 1%. Inaccuracy of the data on the solubility is, in general, not a serious difficulty in the preparation of good-quality crystals. How to avoid this difficulty will be explained in a section on growing crystals by various methods.

If the values of the solubility of a substance in several solvents are known, the solvent should be selected on the basis of the following considerations:

a) very toxic and corrosive solvents should be avoided;
b) very volatile solvents require either careful hermetic sealing of the crystallizer (which may not be easy) or careful control of the rate of evaporation;
c) a dissolved substance should have maximum possible stability, and this also applies to the solvent itself because decomposition products usually have a very strong effect on the growth of a crystal;
d) in some cases considerable difficulties are encountered in the case of solvents in which the solubility of a particular substance depends very strongly on temperature: when crystals are grown using such solvents, the temperature in the crystallizer may have to be maintained rigidly at some constant value;
e) the lower the solubility, the slower (in general) the growth of a crystal, the longer the process, and the greater the chance of failure;
f) if the solubility is very high, solutions, particularly those of organic substances, become so viscous that spontaneous convection in a solution is practically impossible. In the absence of forced mixing such a solution splits into supersaturated regions where spontaneous nucleation may take place, i.e., stray crystals may be formed. Moreover, crystals grow very slowly in viscous solutions and crystals grown in such media usually contain inclusions and exhibit a tendency to grow preferentially at the edges and corners. In other cases, a high concentration of strongly colored compounds in a solution makes it difficult to follow visually the growth of a crystal.

If the solubility is very high and no other suitable solvent is available, the solubility may be reduced by adding a suitably selected liquid which mixes well with the solvent but in which the solute dissolves with difficulty. Thus, the solubility of ionic compounds can be reduced appreciably by the addition of methyl or ethyl alcohols, dioxane, etc. However, we must remember that such solvents are usually unsuitable for the growth of crystals by the evaporation of a solvent since their component parts evaporate at different rates. Moreover, the presence of the additional component in such a solution may have a strong effect on the growth of a crystal, which may not be known in advance.

§20. Preliminary Information on Growth of Crystals

The present state of the theory of the growth of crystals does not permit us to predict the optimum conditions. The great complexity of the crystallization process offers no hope that this problem will be solved in the nearest future. Therefore, it is important to become acquainted with the available theory and to find the optimum conditions by trial and error. It would be desirable to construct an algorithm for the most rational selection of the optimum crystallization conditions. This requires a very thorough knowledge of the laws governing the formation of crystal faces and of defects in crystals; one may hope that crystallographers will be able to develop such algorithms for industrial use.

So far, the problem has not been solved and the following methods for acquiring information on the characteristic features of the growth of crystals from solutions in selected solvents may be suggested. Lack of success in the preparation of large crystals by a selected method frequently requires a new search for a suitable solvent, a study of impurities, and a change of the growth method. This is usually laborious and sometimes the search for the optimum conditions (in the case of growth of large crystals suitable for industrial uses) has taken large teams of investigators many years of work. We must point out that even when large crystals are obtained it is often necessary to search further for better growth conditions.

Familiarity with the crystallization conditions of a particular substance can be acquired as follows.

1. If the volatility of a solvent is considerably higher than that of a solute, then solutions saturated at room temperature or higher temperatures up to 60°C can be prepared in amounts of 4-5 cm^3. If the solubility is unknown, a solution should be kept at a suitable temperature for 1-2 days and then it should be carefully decanted. Then, it should be divided into four approximately equal portions and placed on four watch glasses. Small seeds should be introduced into these portions by using a needle or a similar device. All that is necessary is to bring a seed into contact with the solution. These four portions of the solution are kept at the saturation temperature, and two portions from each group are covered with filter paper. The other pair is covered with a filter paper on top of which is a piece of flat glass in order to slow down the evaporation appreciably. To check that the results are reproducible, this procedure should be repeated at least twice or three times. These watch glasses should not be uncovered, moved, or disturbed in any way before the evaporation is complete, because such action may disturb the growth process. When the solvent has evaporated completely (organic substances are best evaporated under an exhaust hood) the small crystals obtained are examined with binoculars or under a microscope. In general, one should concentrate the attention on individual crystals formed in the center part of the watch glass.

2. If the volatilities of the components of a solution are similar, then solutions should be prepared, in the way just described, at 25 and 60°C in amounts of 10-15 cm^3 and they should be poured into weighing bottles. Once again seeds should be introduced into these solutions and the temperature in the drying chamber, into which these bottles should be placed, should be re-

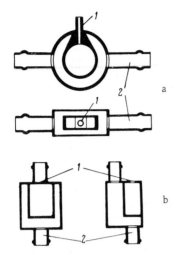

Fig. 46. Cuvettes for the observation of the growth of crystals under a microscope: a) in transmitted light; b) in reflected and transmitted light. 1) Aperture for the introduction of a crystal holder; 2) tubes for the supply of heat exchange agent.

duced by about 5 deg C. After 24 hours, the solution is poured off, and the crystals are transferred onto filter paper and dried rapidly. Crystals grown at high temperatures are best removed from the solution in a heated chamber.

3. If a pure solution is for some reason unsuitable, it is necessary to select an additive (bearing in mind the points mentioned in §5) and to proceed as follows.

An additive is introduced into a series of batches of the main solvent. One must try to use minimum amounts of such an additive because of the possibility of its reaction with a crystal and the effect on its physical properties. As far as the growth is concerned, there is always some optimum (usually very low) concentration of the additive below which the additive is not effective and above which it is harmful.

The substance to be crystallized is added to different portions of such a two-component solvent in such a way that the excess of the solute remains at the bottom of the container for several days (stirring is advisable). This should produce nearly saturated solutions. The subsequent stages in the procedure are the same as those mentioned in 1 and 2 above.

These methods are relatively simple and quite sufficient for the determination of the quality of the solvent but they do not give quantitative data for estimating the necessary supersaturation and temperature during growth, the lack of which may give rise to difficulties in developing large-scale methods.

4. A better method is to observe the crystallization process under a microscope in a thermostated cuvette and record the results by photographing with a still or cine camera [Petrov (1957)]. This approach needs more time but the compensation is that the results obtained are of intrinsic scientific interest. Instead of the special thermostat described in Petrov's paper, one can use commercial-type ultrathermostats in which the liquid is set in motion by some means. These thermostats can be the Soviet TS-15 or TS-24 type, the German U-6 or U-8 type, or Hungarian types of similar construction. The most convenient devices for the study of single crystals growing around a seed are cuvettes in a thermostated jacket, shown in Figs. 46a and 46b. They are attached to the stage of a microscope which can be placed in a vertical position (microscopes of the MP-5, MP-6, and MP-7 type are recommended). The crystal holder may be in the form of a glass rod 0.6-1 mm thick [cf. §28]. In such a cuvette, the saturation temperature can be conveniently determined from the shape of the edges and corners of a crystal, which are the points most sensitive to a transition from growth to dissolution, and conversely.

An occular fitted with a micrometer makes it possible to determine the rate of growth of faces of various forms of a crystal. This is best done by photographing a crystal and then measuring its increase in size per unit time. Cinematography is the most complex but also the most effective method for investigating the kinetics of crystallization. Using the cinematographic technique, one can employ small amounts of a substance and find the optimum crystallization conditions in a relatively short time. One can also determine the range of supersaturations and temperatures within which crystals grow without inclusions and other defects, and it is possible to select a suitable additive and a satisfactory solvent. One must remember that when large crystals are required the range of supersaturation at which growth takes place without inclusions is usually narrower than in the case of small crystals, even when stirring is used.

5. Finally, the most effective method for investigating the growth of crystals is the technique in which a moving solution is used. Several variants of this technique have been described in §10.

§21. Some Methods for Control of Growth Conditions and Crystal Quality

In spite of the great variety of crystal structures and the compositions of systems in which crystallization can take place, as well as the enormous differences between the solubilities of substances, crystal growth obeys certain general relationships. This makes it possible to use model experiments to study the growth of various (including organic) synthetic crystals by investigating mineral-forming processes. It also makes it possible to describe general procedures for avoiding certain typical defects in growing crystals. We shall consider these problems in decreasing order of the frequency of occurrence of a given type of defect. This order is identical with the order of decreasing knowledge about causes of such defects and of increasing difficulty of avoiding them.

1. A growing crystal may have inclusions or it may exhibit skeletal growth at the edges and corners (cf. Figs. 15 and 16). The various forms of these defects and the factors responsible for their appearance are discussed in §§2 and 6. We shall simply give here the measures useful in avoiding such defects.

a) Crystals may be grown under dynamic conditions.

b) The supersaturation can be reduced to avoid these defects.

c) The crystallization temperature can be increased [Petrov (1964)]. It is known that the diffusion coefficient increases by 30-50% for every 10 deg C rise in temperature. Therefore, at higher temperatures, the diffusion is accelerated and this reduces the concentration gradient along a growing face. Moreover, an increase in the temperature of solutions containing substances with a positive solubility gradient may reduce the concentration of the solute in the solution. Perhaps the most important influence of temperature is its effect on the solvation shell surrounding a crystal, which becomes less stable at higher temperatures.

d) The chemical composition of the crystallization medium may be changed by purifying the solvent and the solute, adding a suitable admixture to the main solvent, or using a different solvent. One should proceed in this order, i.e., one should start with purification and end with a change of the solvent. If all these operations do not give positive results, one should try growing crystals from the melt or from the gaseous phase. Usually, this is an extreme measure because it is relatively easy to avoid the formation of inclusions.

2. Anisometric growth (growth at very different rates along different directions) can be only arbitrarily called a defect. Needle-like or platelet crystals can be very homogeneous but sometimes they are of inconvenient shape for further investigations. The problem is how to control the relationship between the rates of growth of different crystallographic faces.

a) Experiments show that a reduction in the supersaturation is usually accompanied by a transition from strongly anisometric to isometric forms of growth. Thus, $KClO_3$, which usually grows in the form of thin plates, forms short prisms at very low supersaturations. Crystals of $MgSO_4 \cdot 7H_2O$ [Mokievskii (1951)], KNO_3 and other substances change from the needle-like shape to isometric forms when the supersaturation is reduced. To compensate for the reduction in the rate of growth at low supersaturations, one should use dynamic growth conditions.

b) The crystallization temperature may be increased because it is one of the factors which affects the shape of crystals and it can give positive results [Mokievskii (1951), Gavrilova (1968)]. This approach has not been investigated much.

c) An extreme but fairly effective method is the change of the chemical nature of the crystallization medium. In this case, one should start by using additives and, if necessary, one can use a different solvent.

d) If all these methods are ineffective, one can use mechanical means to stop crystal growth along some directions; this is particularly suitable for crystals of elongated shape and it is achieved by the use of special crystal holders (cf. §26).

3. Block substructure, nodular growth, and, in the extreme case, the formation of dendrites and spherulites may be observed. Among crystals showing block substructure are those of potassium ferrocyanide (Fig. 47), sodium chloride grown from an aqueous solution, and sorbitol hexaacetate [Gavrilova (1962)]. In crystals without a center of symmetry (quartz, lactose), block formation is accompanied by a characteristic twisting of the crystals.

The causes of block formation in crystals may be heterometry (cf. §5), the attachment of nuclei to a crystal in subparallel positions, and the growth of plastically deformed crystals. In spite of the fairly frequent occurrence of this form of crystal, it has been studied relatively little.

The formation of blocks and the appearance of several protuberances in crystals under given experimental conditions usually begins when crystals reach some definite size, which is typical of a given crystal and particular crystallization conditions. Therefore, the absence of blocks in crystals prepared on a microscopic scale cannot be regarded as a guarantee that they will not appear in large crystals. On the other hand, if small crystals exhibit this effect, it will be definitely present in large crystals and, therefore, we must try to suppress this defect before trying to grow large crystals.

Spherulites (Fig. 48) and dendrites are formed under a great variety of conditions. In most cases, they appear in viscous media, in the presence of strongly adsorbed impurities, or at high supersaturations.

Fig. 47. Block-shaped crystals of potassium ferrocyanide. ×2.

§21] METHODS FOR CONTROL OF GROWTH CONDITIONS AND CRYSTAL QUALITY 75

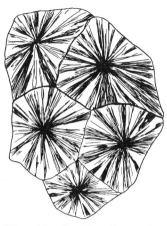

Fig. 48. Section through a crystal of aluminum sulfate showing spherulites.

There are only a few known methods for avoiding the formation of spherulites and dendrites. The methods recommended include:

a) The use of high-quality undeformed seeds.

b) The reduction of the supersaturation.

c) The change of the chemical nature of the medium by purification of the solution or the use of a different solvent.

Thus, it has been reported that the quality of crystals of isatin is improved by the replacement of the lower alcohols, used as solvents, with higher alcohols [Hartmann (1964)].

4. Stresses may appear in crystals, the harmful effects of which range from anomalous birefringence to crack formation. The presence of stresses is a common defect, particularly in large crystals. The way in which stresses manifest themselves depends, primarily, on the magnitude and configuration of the stress field and on the nature of the material: its mechanical and optical properties, plasticity, and strength.

Stresses in small crystals are not common and, therefore, measures to avoid stresses and the formation of cracks are usually necessary only when large crystals are grown. One of the typical causes of the appearance of stresses in crystals is the heterometry. Apart from the causes already referred to, an important source of stresses is the use of a seed grown under conditions differing strongly (in the chemical nature of the solvent, the supersaturation, etc.) from the conditions which apply during growth of the main crystal around a seed (Shternberg (1962)]; also sudden changes in the growth conditions may produce stresses. The main methods for avoiding stresses in crystals are as follows:

a) The preparation of seeds under conditions similar to those which will be used later in growing crystals.

Fig. 49. Porcelain-like crystals of lead nitrate. × 2.

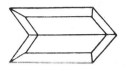

Fig. 50. Gypsum twin.

b) The stabilization of the growth conditions, particularly the maintenance of a constant temperature.

c) The reduction of supersaturation [Gavrilova (1962)]. Gavrilova has not cited heterometry as the cause of stress formation but this can be expected from the fact that an increase in the supersaturation usually increases the anisotropy of the rates of growth and this should tend to produce large differences in the compositions of different growth pyramids.

When the stresses are due to sectorial heterometry, it is necessary to find the impurities whose concentrations vary most widely in different growth pyramids and remove these impurities from the solution.

Another fairly common cause of the appearance of stresses and cracks is the capture of solid particles by a crystal at higher temperatures in a solution. Stresses can be produced when the thermal expansion coefficient of a crystal is considerably greater than the expansion coefficient of a captured particle. A particle captured at a high temperature is subject to a crystallization pressure [see Khaimov-Mal'kov (1960)] and it becomes strongly compressed during the cooling of the crystal because of the difference between the expansion coefficients. If the particle has a low compressibility, the parts of the crystal next to it are pushed apart and cracks may form. This cause of stress formation is more important at high temperatures, and cracks in naturally occurring crystals are due to this cause. Optical anomalies around occluded solid particles are of the same origin. Moreover, cracks may appear around a solid crystal holder when the holder is vibrated for the purpose of stirring (cf. §26).

Visualization of stresses in crystals by optical methods is described by Doladugina and Berezina (1968a and 1968b).

5. The formation of microinclusions and of white porcelain-like crystals is due to the appearance of islands on the surface of a face, which are due solely to the attachment of strongly adsorbed impurities. These defects are found in small crystals. Lead nitrate (Fig. 49)

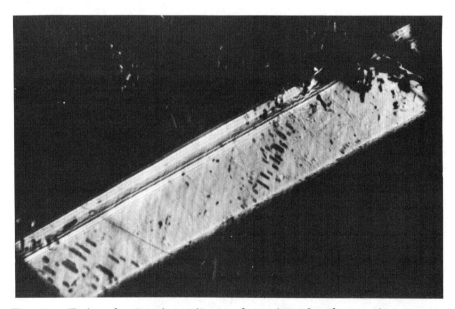

Fig. 51. Twinned potassium nitrate plate viewed under a microscope between crossed Nicol prisms. The long parallel stripes represent the different species of a polysynthetic twin. × 100.

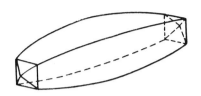

Fig. 52. Curved crystal of ammonium dihydrogen phosphate.

and potassium iodate, grown from aqueous solutions, are the most typical examples of crystals exhibiting this type of defect.

The only known method of avoiding this defect is a change of the chemical nature of the crystallization medium. In particular, positive results are obtained for $Pb(NO_3)_2$ by the addition of HNO_3, and in the case of KIO_3, the addition of HIO_3 is helpful [Kasatkin, Petrov, and Treivus (1962)].

6. Twinning is common in crystals of tartaric acid, gypsum (Fig. 50), potassium iodate, etc. Twinning is easily observed by the presence of re-entrant angles between faces. Frequently, twinning cannot be observed from outside and only an examination under a microscope between crossed Nicol prisms shows that a crystal does not have a uniform extinction coefficient (Fig. 51).

In view of the relative scarcity of information on the causes of twinning, it is difficult to recommend measures for avoiding them. However, it has been reported that at low supersaturations twins are formed less frequently than at high supersaturations [Polukarov and Semenova (1966)].

7. Curved crystal faces are sometimes observed. Such faces frequently grow very slowly. The most typical examples of crystals exhibiting this defect are ammonium and potassium dihydrogen phosphates (Fig. 52).

The main cause of this effect is the specific action of strongly adsorbed impurities on the rate of growth of some faces. In the case of ammonium and potassium dihydrogen phosphates, the cause is the strong leaching of heavy elements (Fe, Ca, etc.) from the walls of glass crystallizers by these two compounds; these heavy elements form compounds with PO_4^{2-}, which are difficult to dissolve and are not isomorphous with the two phosphates. This may result in curved faces. The curved-face effect is found also in small crystals.

There are several methods for avoiding this type of defect:

a) The chemical nature of the solution, particularly its acidity [Byteva (1965)], should be changed.

b) The temperature should be increased [Gavrilova and Kuznetsova (1966)].

c) The supersaturation may be increased (this is not a very effective method) [Gavrilova and Kuznetsova (1966)].

Thus, in general, the homogeneity of a grown crystal is governed by: 1) the absolute values of thermodynamic parameters; 2) the stability of these parameters during crystallization; 3) the gradients of these parameters along the surface of a growing crystal; 4) the quality of the seed crystal; 5) the degree of match between the structure of the seed and the growing crystal.

§22. Selection of Crystal Growth Method

The selection of the growth method is governed by the following factors: A) the nature of the solubility of a given substance; B) the physicochemical properties of the solvent and the solute (volatility, chemical stability, etc.); C) the required dimensions of a uniform crystal, bearing in mind the limitations of the method and the growth characteristics of a particular substance; D) the apparatus available in a given laboratory. At present, there are practically no standard crystal-growing apparatus which can be acquired commercially and, therefore, the last factor is frequently of great importance.

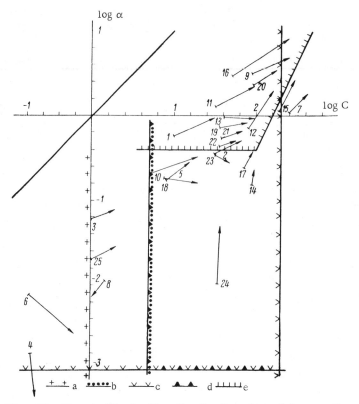

Fig. 53. Graph illustrating the limits of usefulness of various crystal growth methods, shown as a function of the value of the solubility C and of its temperature coefficient α. 1) Potash alum; 2) sodium bromate; 3) hippuric acid; 4) gypsum; 5) potassium iodate; 6) calcium iodate; 7) iodic acid; 8) lithium carbonate; 9) urea; 10) barium nitrate; 11) potassium nitrate; 12) sodium nitrate; 13) lead nitrate; 14) strontium nitrate; 15) sucrose; 16) Rochelle salt; 17) ammonium sulfate; 18) potassium sulfate; 19) potassium hydrogen sulfate; 20) magnesium sulfate heptahydrate; 21) copper sulfate pentahydrate; 22) ammonium chloride; 23) potassium chloride; 24) sodium chloride; 25) lead chloride. a) Crystallization by chemical reactions; b) solvent evaporation method; c) thermal and forced convection methods; d) concentration convection method; e) cooling method. The special marks along the boundary lines (crosses, triangles, dots, etc.) are on the inside of the region in which a given method is applicable.

We shall now consider four factors in turn.

A. Nature of the Solubility and Selection of the Growth Method.
When one speaks of the nature of the solubility and its relationship to the growth method, it is usual to employ terms such as high solubility, poor solubility, strong temperature dependence of the solubility, etc. We have to make these concepts more precise and, therefore, we propose the following classification of substances in accordance with the values and the temperature coefficients of their solubilities.

The proposed classification of the solubilities is as follows: 1) low solubility (less than 1 g per 100 g of the solvent); 2) moderate solubility (from 1 to 5 g per 100 g of the solvent); 3) high solubility (from 5 to 200 g per 100 g of the solvent); 4) very high solubility (more than 200 g per 100 g of the solvent).

The proposed classification in accordance with the temperature coefficients of the solubilities is as follows: a) small temperature coefficient of the solubility: less than 0.001 g/deg per 100 g of the solvent; b) moderate temperature coefficient of the solubility: from 0.001 to 0.5 g/deg per 100 g of the solvent; c) large temperature coefficient of the solubility: more than 0.5 g/deg per 100 g of the solvent.

It is obvious that such classifications are necessarily arbitrary and as the subject develops new classifications may be established.

To select the growth method in accordance with the value and the temperature coefficient of the solubility, we can use the graph shown in Fig. 53. In this graph, logarithmic scales are used to plot the solubility C (along the abscissa) against the temperature coefficient of the solubility α (along the ordinate) for 25 substances grown by the present authors in the temperature range 5-65°C (these substances should be regarded only as examples). The directions of the arrows in Fig. 53 indicate the direction of change in C and α when the temperature is increased. The tail and head of each arrow represents respectively, the values of these two parameters at 5 and 65°C.

The area to the left and above a diagonal line passing through the origin represents the conditions under which the formation of solutions is practically impossible: it represents extremely large values of the temperature coefficient α compared with the values of the solubility C. The other lines in the figure represent limits recommended for a particular growth method on the basis of our experience in growing many substances.

We shall now consider the use of various methods, taking into account the solubility of the substances involved.

If the solubility is less than 1 g/100 g of the solvent (log C < 0), crystals can be grown by the chemical reaction method. The corresponding limit of the range of conditions for this method is denoted in Fig. 53 by a vertical line of crosses (a) on the left; it is based on the experience of growing calcium iodate and lead chloride, as well as certain other substances (fluorite, crocoite, etc.). It is likely that this boundary could be shifted to the right. However, it would be difficult to find suitable reagents which would have a higher solubility than the substance being synthesized. Beginning from solubilities of 5 g/100 g of the solvent (log C > 0.7), we can grow crystals by the solvent evaporation and concentration convection methods (combined vertical line of dots and triangles on the left). At lower values of the solubility, the use of the solvent evaporation and concentration convection methods is possible but not recommended because the growth is very slow and, moreover, in the evaporation method large amounts of the solution would be required and the likelihood of the formation of stray crystals would be high (cf. §13).

The actual value of the solubility is of no great importance in the thermal and forced convection methods, except when solutions become very viscous at very high solubilities. A high viscosity impedes convection in a solution. This is why a vertical line (c) is inserted on the right of Fig. 53 to indicate the limit of recommended use of the thermal and forced convection methods, which should not be employed above 200 g/100 g of the solvent (log C > 2.3). An increase in the viscosity of a solution also impedes the transport of matter during crystallization by the solvent evaporation and concentration convection methods.

The inclined line, which represents the limit of usefulness of the cooling crystallization method, is drawn at solubilities of about 200 g/100 g of the solvent because at these concentrations other methods are preferable (otherwise a large amount of the substance is necessary to prepare a solution). This line is inclined because the higher the solubility the larger are the values of the temperature coefficients of the solubility at which this method is unsuitable. Thus, iodic acid and sucrose should not be grown by the cooling crystallization method but by the solvent evaporation or concentration convection methods.

Depending on the temperature coefficient of the solubility, the regions in which various methods are recommended are distributed as follows.

If the solubility is independent of temperature, we cannot use the cooling method or the thermal and concentration convection methods.

If the temperature coefficient of the solubility is larger than 0.001 g/deg per 100 g of the solvent ($\log \alpha > -3$), we can use the convection methods. Growth is slow but because the number of methods suitable for substances with such temperature coefficients is limited, the convection methods are obligatory.

If the temperature coefficient of the solubility reaches 0.5 g/deg per 100 g of the solvent (sodium bromate, ammonium chloride), we can use — in addition to the convection methods — the crystallization by cooling.

The value of the temperature coefficient of the solubility is not of great importance in the crystallization by chemical reactions or by solvent evaporation. In the case of the concentration convection method, the larger the value of the temperature coefficient of the solubility, the easier is the growth method at low solubilities. Therefore, the upper part of the limit of the usefulness of this method in respect of the solubility should, in principle, be bent to the left. However, the lack of data on the use of this method has prevented us from drawing it more accurately than as shown in Fig. 53.

These limits of the usefulness of the various growth methods are approximate because we can, for example, grow crystals by cooling at gradients lower than those recommended here. In this case, it is necessary to use relatively large amounts of a solution, or to reduce the temperature strongly during growth. Sometimes this has to be done for other reasons. The limits in Fig. 53 simply indicate which method is more economic in materials and time for a crystal of given dimensions although, in principle, large uniform crystals can be obtained by any method.

B. Physicochemical Nature of the Solvent and the Solute. Here, we must consider various properties. For example, the more volatile, corrosive, flammable, and unstable a solution, the lower is the temperature at which crystallization must be carried out. Thus, the cooling and thermal convection methods may be unsuitable when high temperatures cannot be used. When the solvent is toxic and an efficient exhaust system is not available, the solvent evaporation method cannot be used.

C. Characteristic Features of Each Method Considered from the Point of View of the Preparation of Uniform Crystals. Each of the methods described in this book has its advantages and disadvantages in relation to the ease with which the imperfections in crystals can be avoided and in relation to other difficulties arising during growth.

First of all, uniform crystals are more easily obtained by steady-state methods. In this respect, convection methods are most convenient since a crystal is grown at a constant temperature and the impurity concentration during growth is relatively constant. The thermal convection method has the disadvantage of the occurrence of small spontaneous fluctuations of temperature about some average value and the associated fluctuations of the supersaturation. The cooling method is not recommended because the need to reduce the temperature during the process (sometimes by 50-60 deg C) results in a nonuniform distribution of impurities in a crystal, which may produce stresses or even cracks in a crystal. In the solvent evaporation method, the amount of an impurity in a solution also changes during the process.

Second, to obtain uniform crystals preference should be given to those methods in which a relative motion between a crystal and its crystallization medium can be produced easily. In this respect, the thermal and forced convection methods and the crystallization by cooling have advantages over other methods.

Third, the various methods differ in the ease of control of the parameters of the growth process.

Fourth, the efficiency of some methods decreases rapidly when a supersaturated solution becomes unstable. Such instability is not important in the thermal and forced convection methods but it makes it very difficult to prepare crystals using the other methods.

It follows from this discussion that the thermal and forced convection methods and the cooling method are "fast," whereas the other two methods are "slow." Here, "fast" means that the crystallization conditions must be kept constant for days or weeks, and "slow" means that these conditions have to be maintained for weeks or months. Therefore, the "fast" methods are usually employed in the preparation of large crystals.

D. Availability of Crystal-Growing Equipment. In this respect, the crystallization by chemical reactions requires less special apparatus than any other method. The method of solvent evaporation is a little more complicated because of the need to maintain a constant temperature within narrow limits. Then, in increasing order of technical complexity, we have the concentration convection method, the cooling method, and the thermal convection method. Finally, the method of forced convection requires the most complex apparatus. It follows that the more complex methods require better training of the laboratory staff and greater care in their work. Moreover, the more complex the apparatus the longer is the overall time required to grow a crystal, the higher is the probability of some accident, and the more difficult it is to maintain constant conditions.

CHAPTER V

EXPERIMENTAL PROCEDURES

Work in a crystal-growing laboratory is, in many respects, similar to that in an ordinary chemical laboratory. For information on chemical laboratories, the reader is referred to the textbooks by Verkhovskii (1959) and Voskresenskii (1964). Moreover, it is worth reading the booklet "Safety Rules for Chemical Laboratories" (1964). The present chapter will deal specifically with the special procedures used in a crystal-growing laboratory.

It is recommended that a diary be kept in which the date and duration of each operation, crystallization parameters, and the results of observations on growing crystals are methodically recorded. It is important to note in such a diary any differences in a given experiment from previous runs so as to find quickly any relationship between growth conditions and the nature of the crystals obtained. The records should be kept in such a manner as to enable the users to find quickly any required information.

§23. Purification of Reagents

To grow uniform single crystals, we must purify the standard reagents used in chemical laboratories (those of "chemically pure" and "analytical" grades) and this applies even more forcibly to the "pure" and "technical" grades. Reagents of acceptable purity for growth — "spectroscopically pure" and "specially pure" — are available for very few substances and they are expensive.

A reagent may be contaminated with various types of impurity. It may contain mechanical, colloidal, and bacterial particles which are insoluble in a given solvent. It may also contain soluble chemical impurities. The important role of organic dust in the process of nucleation in solutions has been demonstrated in a very clear manner by Rogacheva and Belyustin (1968). Thus, the purification of a substance may consist of the removal of insoluble particles by passing a solution through a paper filter, a porous glass filter, or a ceramic filter (in the case of colloidal and bacterial particles). These filters remove particles larger than 1 μ. Filtration is carried out using the apparatus described in §17.

Harmful soluble impurities are usually removed by recrystallization. Since the segregation coefficients of impurities are not known for the majority of substances and the qualitative influence of impurities present in a solution is not known either, we recommend the following method of preparation of a reagent before recrystallization. The reagent is mixed with a solvent at the highest possible temperature (cf. §24). Then, the resultant solution is overheated (to prevent crystallization during filtration) and filtered. The filtered solution is poured into a crystallizer, which is placed in a thermostat (cf. §15) fitted with a cooling unit in which tap water is circulated. Next, a stirrer is placed in the crystallizer and a cover is fitted on. The stirrer is set in motion. Purification by recrystallization under stirred conditions is particularly effective [Matusevich (1959, 1960)]. A solution should be stirred for 1-2 days. Then, the stirring is stopped, and the solution is shaken to ensure thorough mixing and filtered again. After the filtration, the precipitate remaining in the filter is washed with a small amount of the solvent, which removes mother liquor from the precipitate.

The initial amount of a solution depends on the amount of the substance required. The volume of the solution is calculated on the basis of the amount of the substance which may be precipitated from the solution, taking into account the initial and final temperatures of the solution and the nature of the solubility curve. Obviously, the larger the temperature coefficient of the solubility and the larger the temperature difference, the more effective is the purification. The lower the solubility of a substance at the lowest temperature employed, the more economic is the purification.

After the first recrystallization, the two components (the precipitate and the remaining solution) should be tested by the crystallization of small amounts on watch glasses or in Petri dishes. The quality of the crystals which are grown in this way is used to select the next step in the preparation of a reagent for final crystallization. Usually, the quality of the crystals grown from a solution prepared from the recrystallization precipitate is higher than the quality of crystals grown from the solution remaining after recrystallization (for an impurity segregation coefficient $K<1$). In this case, the working solution is prepared only from the recrystallized precipitate. Sometimes, the distribution coefficient is greater than unity and then the solution obtained after recrystallization is purer than the precipitate left in the filter. In such cases, the precipirate is rejected and only the solution is used.

In many cases, the quality of the final crystals is improved considerably by a second recrystallization. In the simplest case, the precipitate remaining in the filter is redissolved in a new batch of the solvent. The more economic methods of recrystallization used in industry are described by Mullin (1961). We must bear in mind that the smaller the amount of the residual impurity the more difficult it is to remove: each recrystallization is less effective than the preceding one. Moreover, the smaller the crystallochemical differences between impurities and the host substance the less efficient is the recrystallization process. The most difficult case to deal with by recrystallization is a substance containing an isomorphous impurity, such as potash alum containing an admixture of chromium alum.

The residual solution, enriched with the impurity, is simply thrown away or, if the substance is expensive, the solvent is evaporated and the reagent is retained. Such a reagent has a higher impurity concentration than it had before crystallization, and several more purification stages would be needed before it could be used again.

It is worth spending much time on purification only if it is known that the impurity present in a reagent has a harmful effect on the growth, because purification processes are laborious and result in a considerable loss of the reagent.

Solvents are usually purified by simple distillation. If a solvent has to be particularly pure, it should be subjected to rectification [Krell (1957)].

Volatile substances, for example, acetoxime, are best purified by sublimation. The temperature should be controlled in order to avoid the decomposition of the reagent.

Impurities can be removed also by chemical means. In this case impurities are precipitated by means of those reagents which are normally used to detect these impurities by qualitative analysis. If these impurities originate from the crystallization apparatus itself, they can be made inactive by using complex-forming agents which produce strongly bound compounds with impurities and thus prevent them from affecting the growth of crystals.

However, the most effective method for the purification of a substance is the zone melting method. It is widely used to prepare very pure materials in which the impurity concentration represents a very small fraction of 1% [Pfann (1958), Herrington (1963)]. The principles of zone melting are as follows.

The substance is placed in a tube, melted, and crystallized. This produces an ingot. The lower part of the tube is placed in a short cylindrical furnace. The substance in the tube melts only in the region enclosed by the furnace. The tube is slowly lowered and the molten zone moves slowly upward while the substance below it crystallizes again. This pushes the impurities in the upward direction. When a high purity is required, the process is repeated several times. The rate at which the tube is lowered, the temperature to which a substance is heated, and the cooling conditions are selected by trial and error. The purest materials are obtained by a floating zone variant of the zone refining method. In this case, a rod of the substance to be purified passes through a narrow high-temperature zone. The rod melts in this zone but the melt is held in place by surface tension. This method avoids possible contamination from the walls of the tube because the rod is not in contact with it.

Other methods of purification, for example, ion exchange methods, are described in the chemical literature [Yur'ev (1964), "Ion Exchange and Its Applications" (1959), Griessbach (1957)].

If for some reason purification is not feasible in a crystal-growing laboratory, one should purchase very pure reagents from organizations or from suitable chemical institutes or factories (in the USSR these are, for example, the "Red Chemist" factory in Leningrad, the Applied Chemistry Institute in Leningrad, and the Chemical Engineering Institute in Moscow).

§24. Preparation of Solutions

To prepare a solution, it is necessary to know the solubility of a substance in grams per 100 g of the selected solvent. Since the expressions used by different authors to calculate the concentration differ considerably, it is useful to know the various ways of expressing concentrations so as to be able to convert them to the required form. The more usual forms of expressing concentrations are:

a) Weight percent (the percentage of a substance in a solution if the weight of a solution is assumed to represent 100%).

b) Molar percent (the number of gram-molecules or moles of a substance contained in a solution, expressed as a percentage of the total number of moles).

c) Moles per 1000 g of the selected solvent (molal scale) and moles per 1000 cm^3 of the solution (molar scale).

It is not difficult to convert the concentration from one form to another. Only in the case of conversion from and into the molar scale do we need to know the density of the solution (usually the results given on the molar scale are given together with the density of the solution).

The values given in various reference books usually refer to the solubility of anhydrous substances. Reagents are usually obtainable in the form of hydrates. Therefore, in determining the amount of a substance necessary to prepare a given solution we must take into account the presence of the water of crystallization. The conversion of the amount of anhydrous salt to an equivalent amount of a hydrated salt is carried out using the formula

$$X = \frac{100 M_h C}{100 M_a - 18 n C} \text{ g of hydrate}/100 \text{ g of solvent},$$

where M_h is the molecular weight of a hydrate; C is the solubility of an anhydrate, expressed in g per 100 g of water; M_a is the molecular weight of the anhydrate; n is the number of water molecules in the formula for the salt in question.

The derivation of this formula is elementary. It is useful to plot and keep handy a solubility graph. In the absence of data on the solubility, it is necessary to determine approximately the solubility using the methods described in §25, or, more exactly, by observing the behavior of a crystal in a solution (cf. §26).

Solutions should be prepared as follows.

1. First, the necessary amount of the solvent is measured. Its volume should be measured with an accuracy of the order of 0.5-1%. In the absence of an accurate measuring cylinder, the solvent must be weighted. Pipettes should be used to add or remove small amounts of the liquid.

Then, the solvent is heated in a flask to a temperature 10-20 deg higher than the saturation temperature in order to accelerate the process of dissolution. Excessive overheating should be avoided because it results in an appreciable loss of the solvent.

2. A pure solute is ground to a powder of grain size not larger than 1 mm; the necessary amount is weighed and added to the hot solvent. The weighing error must not exceed 0.5%.

3. Vigorous shaking is used to mix the content of the flask until the solute dissolves completely. A flask containing a solute at the bottom should not be left on a heating plate for a very long time because it may crack. The flask should not be closed by a stopper because when the solution boils due to overheating, the stopper and the solution may be ejected from the flask. It is best to place a vertical condenser in the stopper and to circulate water through the condenser. In this way, the evaporating solvent is condensed back into the flask and there are practically no losses of the solvent; at the same time, the danger of the stopper's being pushed out is avoided. The outlet of the inner condenser-tube is stopped with cotton wool.

4. The solution must not be allowed to cool before filtration. After filtration, the solution is ready for the next operation. While waiting for the next stage, the neck of the flask should be covered with filter paper or stopped with cotton wool to prevent dust and other particles present in the atmosphere from contaminating the solution. Contamination with crystalline dust would later result in the appearance of stray crystals; biological dust may produce the same effect, and give rise to fungus colonies in the solution. This happens frequently in the case of solutions of alums, asparagine, glucose, etc. If it is essential to carry out work under sterile conditions, specially made Plexiglas (polymethylmethacrylate) containers should be used.

When no data are available on the solubility of a substance, it is best to prepare a solution by mixing. The concentration corresponding to saturation at a given temperature can, in principle, by reached in two ways: by dissolving a substance at a selected temperature or by precipitating an excess of the substance from a supersaturated solution. The former method is usually employed because the process of dissolution is usually faster than the crystallization. The difference between the crystallization and dissolution rates (noted many years ago) can sometimes be very large [Slavnova, Gordeeva, and Sitnik (1965)].

Mixing is a very simple process. A substance is placed in a crystallizer (Fig. 34), the selected solvent is poured in, and a stirrer is placed at a position which is at a quarter of the height of the solution column. The stirrer is made, for example, from a glass rod and a piece of Teflon sheet. Such a rod has two thicker parts separated by 3-5 mm. A hole, equal to the diameter of the glass rod, is made in the Teflon sheet (1-2 mm thick). One of the thicker parts of the rod is forced through the hole in the Teflon sheet, which is then held between the two thicker parts of the rod.

A reciprocating motion (of suitable amplitude and frequency) of the stirrer mixes the solution and accelerates the saturation stage. Usually 24 hours are sufficient to produce a state of equilibrium between the solution and the solid; in some cases, two days are needed.

Then, the stirring is stopped and the solution is rested for 1-2 h. If a recrystallized substance has been used to prepare the solution, no filtration is necessary. However, if a crystal is to be grown by cooling, the crystallizer should have the same dimensions as the container in which the mixing was carried out. The crystallizer should be heated first in a dessicator to a temperature approximately equal to the mixing temperature. The container with the solution is taken out of its thermostat and the empty crystallizer is placed immediately in the thermostat. The solution is poured into the crystallizer carefully so as not to disturb the precipitate at the bottom of the container. Next, if high accuracy is required, the saturation temperature is determined and the solution is heated by 3-4 deg above this temperature before introducing a test seed (cf. §26). Experience in this operation can be acquired in several runs.

§25. Determination of Solubility

In spite of the extensive literature on the solubility of various substances, it is frequently found that the necessary data on a given substance are lacking. Consequently, it is necessary to find a suitable solvent and sometimes to determine the solubility of the reagent to be used. There is as yet no general theory of solubility and, therefore, it is not possible to predict how well a given substance will dissolve in a given solvent. However, there are some relationships which help in this matter but unfortunately exceptions to these general rules are many. In some cases, it is simply necessary to try all the liquids available in a laboratory.

The basic rule, established centuries back by alchemists, says that the solute and the solvent should be similar. The similarity applies to the degree of polarity of the interacting substances (this interpretation of the rule is, of course, of later origin). Thus, polar and ionic substances dissolve easily in polar solvents. Nonpolar substances dissolve more easily in nonpolar solvents. Some information on the selection of a solvent for organic solids can be found in the books by Yur'ev (1964) and Weissberger et al. (1955).

In the search for a suitable solvent and in the prediction of its properties, we can also use two rules which follow from Schröder's solubility equation [Kirillin and Sheindlin (1956), p. 69]. These rules relate the solubility of a substance to its melting point and the latent heat of fusion:

a) At a given temperature, a solid with a higher melting point dissolves less easily in a particular liquid than a solid with a lower melting point.

b) When two solids, having different melting points, are dissolved in a particular liquid at a particular temperature, the less soluble is that solid whose latent heat of fusion is larger.

A rough estimate of the solubility can always be obtained by the investigator and such an estimate is usually sufficient for the selection of the growth method. In general, in order to save time, the solubility can be determined simultaneously for several liquids.

Between 2 and 3 cm^3 of a solvent are poured into a test tube and the investigated solute is placed in the tube in an amount representing about 20% of the volume of the solvent. Observations are made of the process of dissolution. If shaking makes the substance dissolve completely, more solute is added and the solution with a precipitate is stored in a closed test tube for several days at a constant temperature; it is desirable to apply some form of stirring or mixing during this period. After being cooled, the solution is carefully decanted into a weighing bottle in such a way as not to disturb the residue at the bottom. After being cooled, the weighing bottle is weighed (P_1). If the volatility of the solute is considerably less than the volatility of the solvent, the solvent is evaporated (this can be done slowly in air under an exhaust hood).

The degree of dryness of the residue is checked by periodic weighing. The constancy of the weight (P_2) in two consecutive measurements indicates that the drying is complete. The weighing bottle is washed out and the dry bottle is weighed once again (P_3). Then, $P_1 - P_2$ is the weight of the solvent (P_S), where $P_2 - P_3$ is the weight of the solute (P_A). Hence, the solubility, expressed in grams per 100 g of the solvent, can be calculated from

$$c = \frac{P_A}{P_S} 100.$$

If the volatilities of the solvent and solute do not differ greatly, more complex procedures have to be followed. In particular, it is necessary to weigh separately the solvent and the solute. The solvent and the solute are mixed, the mixture is made homogeneous by increasing its temperature, and then the saturation temperature is determined by one of the methods described in §26. Determination of the saturation temperatures of a series of solutions of various concentrations makes it possible to determine the solubility curve.

§26. Determination of Saturation Temperatures of Solutions

To achieve success in some of the crystal growth methods, it is necessary to determine the saturation temperature of a solution. The concentration of a prepared solution (§12) is usually only very roughly equal to the necessary saturation concentration.

The saturation temperature can be determined in several ways. We shall describe in detail the following methods: 1) by the observation of convection currents; 2) by the observation of the growth and dissolution of crystals; 3) by measuring the electrical conductivity of the solution.

Determination of the Saturation Temperature by the Observation of Convection Currents

We have already mentioned (cf. §6) that when a crystal is grown in a fixed position, convection currents rise upward from it. When a crystal is dissolved, such currents proceed in the downward direction. Under equilibrium conditions, these currents are absent. Thus, by observing convection currents, we can determine the saturation temperature of a solution. The procedure is as follows.

a) A uniform crystal of 5-10 mm diameter is used. If a nonuniform crystal (one containing inclusion) is employed, then small fragments may break away from it. These fragments can drop to the bottom of the container and give rise to stray crystals. If a crystal smaller than the indicated size is used, convection currents may be difficult to observe.

b) The crystal is attached to the end of a chemically inert wire, 0.7-1 mm thick, using an adhesive which does not dissolve in the solvent being investigated. For this purpose, a small cavity is drilled carefully in the crystal to a depth of 3-4 mm. This cavity is filled with an adhesive and the end of a wire or a rod (which is first degreased in acetone) is inserted into the cavity. When the adhesive dries out, the test crystal is ready for use. If a wire or a thread is wound around a test crystal, the crystal frequently falls away from its support. Consequently, in such a case the experiment has to be stopped, the solution heated again, etc. In the case of alums, Rochelle salt, and other substances with low melting points, we can use a simpler method for attaching a test crystal to its support. In such cases, the end of an enameled copper wire (0.5-1.0 mm thick) is cleaned so as to remove the enamel over a length of 5-8 mm. Using a spirit lamp, the clean end of the wire is heated and the hot wire is pushed into a crystal to a depth of 3-4 mm. It is necessary to hold the wire and the crystal together for a few seconds until the molten material around the wire recrystallizes again. After the crystal is attached, the wire is bent near the crystal at right angles to it so as to avoid any interference of the wire with the rising convection currents.

c) The crystallizer is covered with thick glass. The glass cover should have two apertures: one is used to introduce a thermometer and the other to hold the wire supporting the test crystal. The thermometer and the crystal holder are fitted into the cover by rubber stoppers. The mercury bulb of the thermometer and the test crystal should be at approximately the same level in the solution, somewhere in the middle of the volume of the solution.

d) The cover, the thermometer, and the crystal holder are washed in hot (70-80°C) water and the test crystal is rinsed for 3-5 sec in a warm solvent. The solvent is, of course, the same as that being investigated. Next, the crystallizer, heated to 3-5 deg above the expected saturation temperature, is covered and the observations are started.

Convection currents do not start immediately but after an interval of 1-2 min from the moment of immersion of the test crystal in the solution.

e) The observations are carried out using a small light source placed at some distance from the solution. The currents are found near the top and the bottom of the crystal. Sometimes it is difficult to observe convection currents or to distinguish them from flaws in the glass wall of the crystallizer. If there is any difficulty, the crystal or the crystallizer should be slightly turned so as to start the convection currents. If such currents proceed downward, it indicates that the test crystal is dissolving. The temperature and time of such an observation are recorded. Then, the temperature is lowered by an amount which depends on the form of the convection current. If the downward convection current is clearly visible and the diameter of the current is several millimeters, the temperature should be reduced by at least 2-3 deg C. If the diameter of the convection current is small and the current is barely visible, it follows that the solution is close to the saturation temperature and the temperature should be reduced slowly. After each reduction of the temperature, the solution must be allowed to reach a new equilibrium and the temperature and the time of observation should be recorded again. As the temperature is reduced, the downward convection current should become weaker and weaker and finally it should disappear at some particular temperature. This should be recorded. The temperature should be decreased still further but this should be done slowly. Next, an upward current should be observed. Having determined the temperatures at which the upward and downward convection currents stopped flowing (this should be done with an accuracy of 0.1-0.2 deg C), we obtain a narrow range of temperatures in which the solution is saturated.

Shternberg (1944) has pointed out that after the disappearance of the convection currents proceeding from a crystal, a colored halo may be visible near the saturation point: at the beginning of the dissolution of the crystal such a halo should be observed below the crystal, and at the beginning of growth of the crystal the halo should be above it.

Initially, such a process of determination of the saturation temperature from the concentration convection currents is a fairly long process. However, experience of working with a given substance should make it possible to reduce the time taken by such an experiment to 1-1.5 h.

The method described can be used to determine the saturation temperatures of substances with a strong temperature de-

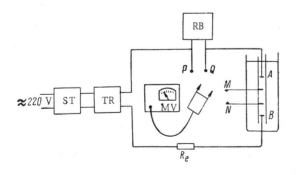

Fig. 54. Circuit for measuring the electrical conductivity of a solution using a four-electrode method.

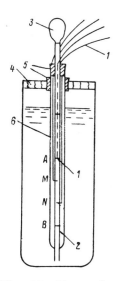

Fig. 55. Four-electrode cell for measuring the electrical conductivity of liquids.

pendence of the solubility, i.e., substances which are usually grown by cooling. The sensitivity of the method can be improved by using raster (scanning) optics [Valyus (1949), Kovalevskii (1958)].

Determination of the Saturation Temperature by the Observation of the Growth and Dissolution of a Crystal

This method is technically a little more difficult than the convection current technique. To use it, we must have a thermostated cuvette, which is viewed using binoculars or a microscope (cf. §20).

To determine the saturation temperature, a portion of a prepared solution is removed from the crystallizer by means of a pipette and placed in the cuvette. The pipette must be dry and heated before use. The solution in the crystallizer must be heated to a temperature 5-10 deg C higher than the expected saturated temperature. By varying the temperature in the cuvette, we can change the conditions from growth to dissolution and conversely. When a crystal grows, its edges are straight, its faces reflect light, and sometimes growth centers are visible; when a crystal dissolves, etch figures appear, the edges become rounded, the faces curved, and the general shape of the crystal becomes more rounded. Sometimes motion of whole layers can be observed: the formation of layers on a crystal during growth and the disappearance of layers during solution. This method of determination of the saturation temperature is the most sensitive of the three techniques described here, and it gives the saturation temperature to within ±0.05 deg C. The time taken to determine the saturation temperature by this method can be made shorter, after some experience with this technique, than the measurement by the convection current method.

Determination of the Saturation Temperature by Measuring the Electrical Conductivity

It is known that the electrical resistivity of solutions is a function of their composition and temperature. This makes it possible to determine the concentration and, therefore, the saturation temperature of a solution.

Such measurements can be carried out using an arrangement with four electrodes [K'uang Hang-Hang (1966)]; it consists of two circuits (Fig. 54): a current circuit AB, which is used to pass a current i through the solution, and a measuring circuit MN used to determine the voltage U_{MN} between any two points MN in a solution. The resistivity of the solution is determined using the formula:

$$\rho = K \frac{U_{MN}}{i},$$

where K is a geometrical coefficient, which depends on the shape and dimensions of the electrode cell, the shape and dimensions of the electrodes, and the distance between the electrodes (Fig. 55).

Such a measuring cell (Fig. 55) can be easily made in the laboratory. Electrodes 1 are platinum wires sealed into a glass tube 2 and forming four rings A, M, N, and B in the tube. The distances between the rings are about 2 cm. A rubber bulb 3 is attached to the top of the tube in order to draw into the cell a portion of the solution used in the measurements. The measuring cell is placed in a glass jacket 6 and fitted into the cover of a crystallizer 4 by mean of a rubber stopper 5.

The measurement is carried out as follows. A resistance box (RB in Fig. 54) is used to select the resistance R_{PQ} in such a way that the voltage U_{PQ} is equal to the voltage U_{MN} in the solution. Then the resistance of the solution is

$$R_{MN} = R_{PQ}.$$

A millivoltmeter (MV in Fig. 54) of the V3-3 type is used to measure the voltage; its input resistance is 1 MΩ. To ensure a high accuracy in the measurement of the resistance, it is recommended that all measurements be carried out using such values of U_{MN} that the pointer of the millivoltmeter is near the end of its scale. The position of the pointer can be adjusted by altering the voltage supplied by a transformer TR connected to the power supply via a stabilizer ST.

The geometrical coefficient of the cell K_g is determined in advance of one of the following methods.

1. The coefficient can be found approximately from the formula

$$K_g = \frac{\rho}{R_{MN}},$$

where ρ is the electrical resistivity of water at a particular temperature; R_{MN} is the resistance of water as measured with this cell.

2. The coefficient can be found approximately by calculating the resistance of the cell, regarded as a liner conductor, using the expression:

$$K_g = \frac{V}{(MN)^2},$$

where V is the volume of water in the cell between M and N; MN is the distance between the two platinum ring electrodes.

3. The most accurate method of finding the geometrical coefficient is to measure the electrical conductivity of a standard solution of potassium chloride:

$$K_g = \frac{1}{\chi R_{MN}}; \quad \left(\chi = \frac{1}{\rho'}\right),$$

where χ is the conductivity of the solution, which can be found in appropriate reference works.

A warning must be given that the determination of the saturation temperature of a solution from its electrical conductivity requires a considerable time during the preparatory stage, which involves plotting the temperature and concentration dependences of the resistivity of a solution for each solute. However, if it is planned to carry out many investigations or to grow many crystals, particularly by the cooling method, such preliminary work is fully justified by the subsequent speed and accuracy of the determination of the saturation temperature.

K'uang Hang-Hang (1966) gives the values of the electrical conductivity of potash alum in a wide range of temperatures. A more sensitive but more complex apparatus has been used by Torgesen, Horton, and Saylor (1963).

§27. Preparation of Seed Crystals

A seed is any fragment of a crystal or a whole crystal which is used to start the growth of a larger crystal in a solution. One must not confuse a seed with a nucleus, because the latter is a spontaneous or an accidental crystallization center actually formed or present in a solution.

The maximum size of a seed cannot be specified *a priori* and, in principle, a crystal of any shape or size can be regarded as a seed if it is used to grow a larger crystal. The smallest size of a seed depends on the convenience of mounting it in a crystal holder and on the experimental conditions. A seed used to prepare a large uniform crystal must satisfy the following requirements:

1. It should be a single crystal free of cracks and block boundaries.

2. It should be free of inclusions.

3. Its surface should be free of sharp cleaved edges.

4. It should, whenever possible, be grown under the same conditions as those to be used in the main experiment.

5. It should be of the minimum size compatible with other requirements.

The first requirement is dictated by the observation that defects in a seed are transferred to a growing crystal. Moreover, the presence of block and other boundaries in a seed gives rise to strong stresses because of the crystallization pressure on the seed and such stresses may give rise to cracks and inclusions. In some cases, it is necessary to carry out repeated crystallization in order to obtain a satisfactory seed. Sometimes one has to begin with an intergrowth of several crystals (Shternberg (1961)]. In this case, an intergrowth is enlarged in size by further crystallization, and the best parts are selected and increased in size by further crystallization until a sufficiently perfect single-crystal unit is obtained which can grow to the required size. The second requirement is not as important as the first but in many cases the presence of inclusions of the mother liquor in a seed reduces the uniformity of a crystal grown from such a seed. It was mentioned in Chap. I that dislocations radiate from inclusions and they penetrate a growing crystal.

The third requirement must be satisfied in those growth methods when there is a possibility of the formation of stray crystals. This is because a seed placed in an unsaturated solution may shed fragments which can then give rise to stray crystals.

The fourth requirement is justified in §21.

The fifth requirement follows from the fact that, at the beginning of the growth of a crystal ("regeneration" of a seed), inclusions of a solution are unavoidably formed around a seed (they can form a "negative crystal"). Therefore, the larger the seed the larger is the amount of poor-quality material captured by the growing crystal. It is best to use seeds whose size is not more than 3-5 mm. Such seeds are frequently prepared by simple splitting of larger crystals but this can produce stresses, dislocations, and cracks in a seed. Therefore, to obtain more uniform seeds large pieces should be used and these should then be reduced to the required size by dissolving. Dissolving (and not machining) relieves the stresses and removes microscopic surface cracks and other defects. This is best carried out by careful polishing with a moist and soft cloth (such as chamois leather). If a seed is prepared in this way, the preparation takes a long time, a large part of a crystal is lost and it is difficult to retain the necessary orientation. Therefore, the following more economic method is recommended for the preparation of seeds.

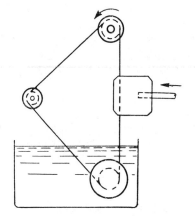

Fig. 56. A device used for cutting crystals with a wet thread. The pulley immersed in the solution hangs freely to maintain a constant tension in the thread. The pulley on the left stretches the thread so that it should not interfere with the motion of a crystal.

A crystal is cut into plates, which are then cut again into small rods of required orientation. The cutting of easily soluble crystals is done by a silk thread wetted with water. For this purpose, one can easily make a small device which is shown schematically in Fig. 56. To cut a crystal into several parallel pieces, three pulleys (two fixed and one hanging freely in a bath with a solvent) are used. If it is desirable to cut a crystal simultaneously into several pieces, each pulley can carry several silk threads in separate grooves. The ends of the threads are joined by untwisting over a length of 2-3 cm, removing some of the strands, and sticking the two ends together with BF-4 or No. 88 adhesive. We must make sure that at the point where the thread is joined the thickness is the same as elsewhere along the length. Crystals can be cut manually by stretching a cotton or a silk thread on the frame of a fretsaw. During the cutting with such a saw, the thread is wetted with a solvent. With this type of thread saw it is possible to prepare, with some care, very small pieces. The cut surfaces, whether produced manually or mechanically, are always uneven and they should be polished by the method described earlier.

If crystals of a suitable size are not available, it is necessary to carry out a crystallization in order to produce seeds. Usually, such crystallization is carried out by cooling a solution. The only difference between such crystallization and the growth of large crystals is that one must try to induce the formation of nuclei. For this reason, a solution is not covered; dust particles fall into the solution and they form nuclei. Usually, small crystals appear at the bottom of a crystallizer after 24 hours. If the nucleation is so effective that a solid lump appears at the bottom of the crystallizer, the solution should be decanted into a different crystallizer (without disturbing the lump at the bottom) and this crystallizer should be covered. In this case, small crystals are usually formed within 24 hours.

If seed crystals are strongly anisometric, i.e., if they are in the form of needles or plates, and if for some reason their shape cannot be changed to isometric by the methods described in §21, satisfactory results can be obtained by enlarging a seed through subsequent growth. To do this, we select a group of the most uniform and large crystals. One of the methods described in the present section is employed to grow them to a larger size. Next, the best and largest crystals are selected from the new batch and these are used in the next stage of enlargement of seeds. Obviously, complete crystals need not be used for this purpose. All we need is the most uniform and largest parts of anisometric crystals, which are best used in the form of plates perpendicular to the long or flat direction of a crystal. This method is labor- and time-consuming but it is frequently essential and it is quite reliable. Dynamic methods, i.e., fast growing methods, can be used in this way to prepare seed crystals of ammonium dihydrogen phosphate, tartaric acid, and guanadine aluminum sulfate.

A different method for preparing thicker seed crystals has been described by Shternberg (1961).

§28. Crystal Holders and Seed-Mounting Methods

The simple growth of crystals at the bottom of a crystallizer is rarely used for a number of reasons. These include the possibility of the dissolution of a seed in an unsaturated

Fig. 57. Crystal holder used to grow crystals under dynamic conditions: 1) rod; 2) polyethylene or rubber tube; 3) seed.

solution; the possibility of contact with stray crystals; and the difficulty of supplying the solute to a growing seed. In order to ensure the best growth conditions, it is necessary to use special crystal holders and the success of an experiment may depend on the suitability of the holder. The selection of a crystal holder and the method for attaching a seed to it are no less important than the selection of the growth method. A crystal holder should ensure that a seed is held securely in the required position and that a seed (and, therefore, a growing crystal) can be moved in any required manner.

A crystal holder should ensure that a seed is not damaged or otherwise adversely affected when the growth begins in a crystallizer. A crystal holder should not become deformed by the selected type and velocity of motion, or by the weight of the final crystal to be grown on it. Otherwise, a gap may form between the crystal and the crystal holder if the latter deforms elastically at a periodic rate. This gap becomes filled periodically and cracks may form when a holder bends back to close the gap. Usually, such cracks are accompanied by many inclusions and the part of the crystal surrounding the holder may be quite useless for further investigations. On the other hand, even in the case of a perfectly rigid crystal holder, the crystallization pressure may produce stresses in a crystal and zones of anomalous birefringence. For this reason, it is best to avoid rigid contact between the crystal and the material of the crystal holder. To achieve this, the crystal holder is covered with a film of elastic lacquer or, which is more convenient, it is isolated from a growing crystal by polyvinyl chloride, polyethylene, or rubber tubing. The end of such a tube can also serve as a seed holder (Fig. 57). Such crystal holders can be used in rotational and vibrational motion of a crystal. A thickening at the end of the holder rod helps to ensure a strong contact with the crystal growing around it, which is particularly important in reciprocating motion.

Experiments have shown that polyvinyl chloride tubes, which are used for electrical insulation, are very suitable for holding a seed in place. If such a tube is immersed for 1-2 min in acetone, it acquires temporary elasticity, which makes it possible to stretch it onto a crystal holder of diameter considerably greater than the diameter of the tube.

If the seed is smaller than the diameter of such a tube, it can be attached to the tube by means of some inert adhesive. BF-2, BF-4, No. 88 or celluloid lacquer adhesives are suitable. We must now consider the method of attachment of a seed to a tube. The seed should be about twice as long as the diameter of the tube. It should be forced into a tube until it is brought into contact with the holder rod because this reduces the possibility of motion of the seed with respect to the holder. The growing crystal should cover the thicker end of the holder and become attached to it as soon as possible so as to avoid relative motion with respect to the rigid part of the holder due to the inertia of the crystal and the hydrodynamic resistance of the solution. Such relative motion, like the deformation of the crystal holder, produces cracks in a growing crystal.

The elastic tube not only acts as the seed holder but also prevents dissolution of the seed if the solution is not saturated. The crystal holder material should be chemically inert in the solution of the substance being crystallized. It should not be dissolved in such a solution even in the slightest degree, nor should it swell, because this may crack the crystal. Most crystal holders are made of glass. In the majority of inorganic solutions, Plexiglas (polymethylmethacrylate) and similar materials can be used. Crystal holders made of these materials

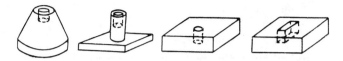

Fig. 58. Crystal holders for locating a seed
at the bottom of a crystallizer.

should be polished in order to prevent the formation of stray crystals in any cavities on their surfaces. In the case of organic solvents, the only reliable crystal-holder material is Teflon (polytetrafluoroethylene). The use of other materials such as Plexiglas, polyvinylchloride, etc., requires a careful preliminary check that there is no interaction between these materials and the solvent (such interactions include swelling, leaching out, dissolution, etc.). Metal rods should be used with care, not only because of their chemical instability but because their high thermal conductivity may disturb the constant temperature conditions near a crystal.

In view of the great variety of possible crystal-holder shapes, we shall mention only typical examples which satisfy, at least to a considerable degree, all the requirements.

1. Truncated cones or pyramids with cavities at the top for the insertion of a seed can be used. One can also use glass tubes or plates shaped as shown in Fig. 58. A glass tube is attached to a plate by a silicate (or some other sufficiently inert) adhesive. Paraffin is poured into the tube or its lower part is closed with a stopper so that the crystal lies in a cavity of 3-5 mm depth. A crystal holder with a seed in it is rinsed in warm water and warm tweezers are used to place the holder carefully at the bottom of the crystallizer.

If a crystal holder is placed in an unsaturated solution, a seed may be partly dissolved and saturated solution may be formed in a cavity produced in this way in the seed. Under these conditions, further dissolution of the seed slows down appreciably if the solution is not stirred. Only a slow molecular diffusion exchange takes place between the saturated solution in the cavity and the undersaturated solution in other parts of the crystallizer. When the temperature is reduced below the saturation value, the seed crystal begins to grow at first solely by molecular diffusion. Next, convection currents appear near the tube carrying the seed and these currents ensure a more rapid supply of the solute and a faster growth of the seed.

It follows that this method of protection of a seed is not absolutely reliable. If the solubility of a given substance depends strongly on temperature, the seed is small, the cavity is shallow, and the overheating is strong or prolonged, the seed crystal may dissolve completely. On the other hand, if the seed is placed too far down in the tube, it will take a long time before it grows out of the tube. To avoid this, it is best to use large seeds. However, large seeds are more difficult to prepare so as to satisfy the requirements mentioned in §27. Preliminary experiments should be carried out to determine the best dimensions of the crystal holder and the optimum experimental conditions.

The crystal holders just described can be used in static, convection, and dynamic methods of crystal growth: a crystal is fixed and the solution is mixed with a stirrer. In this case the crystal holder should be fairly heavy so as to prevent its sliding along the bottom during stirring.

2. To obtain relatively isometric crystals of those substances which normally grow in platelet form, we can use a seed-holder plate with several cavities arranged in a chessboard pattern. Such a plate is placed in a crystallizer and the seeds are set in the cavities either vertically or slightly inclined. The orientation of the seeds is selected to ensure that the plane parallel to the direction of the thickening of the seeds is vertical. In this case, the convection currents ensure a satisfactory supply of the solute to the faces growing at the slowest rate,

which does not occur when flat crystals are placed horizontally at the bottom of the crystallizer. This method of placing the seeds also reduces the possibility of the growing crystals being damaged by stray crystals falling on them.

3. Crystals which normally grow in the form of needles and elongated prisms can be obtained in a more or less isometric form between two parallel plates, which prevent the growth of these crystals along their length.

4. Very small seeds can be held in place as follows. A thin glass-rod holder of about 1 mm diameter is smeared with a thin layer of vaseline. The rod is immersed in a celluloid lacquer or in a cement used for joining cinefilm and magnetic tape; it should be immersed to a depth of 6-8 mm and then withdrawn slowly at a uniform rate. After drying for 5 min, the rod is again immersed in the lacquer or cement for 5 min, withdrawn, and dried for half an hour. If the film prepared in this way is too thin, the operation can be repeated again. After the drying is complete, we are left with a tubular film on a glass rod. The two ends of the lacquer or cement tube are cut with a razor blade and the tube is pulled off the glass rod to a length of 1-1.5 mm so that a seed can be placed in the open end of this tube. A seed crystal whose cross section is approximately equal to the diameter of the glass rod is then introduced into the projecting end of the tube. Light smearing of the ends of the tube with a piece of cotton wool soaked in acetone softens the tube and ensures that the tube firmly clamps the seed crystal and is held securely by the glass rod. After drying in air for half an hour, the crystal holder is ready for use.

§29. How to Treat a Grown Crystal

The extraction of a crystal from its mother liquor requires some care because any damage may destroy completely the scientific value of the crystal or even fracture it altogether.

If a crystal is extracted from a solution kept close to room temperature, it can be simply dried by means of filter paper. Filter paper must not be used to rub the surface since the majority of crystals prepared from low-temperature solutions are easily scratched. The surface of a carelessly treated crystal may acquire many defects. Naturally, if a crystal has to be cut and if its surface is of no interest, these precautions are unnecessary.

If a crystal is grown in a solution kept at a temperature much higher than room temperature (particularly if it is a large crystal whose cross section is more than 1 cm), it may crack because of the rapid cooling when it is exposed to air. To prevent this, a filter paper and a towel are heated approximately to the temperature of the solution. Immediately after its extraction from the crystallizer, the crystal is wrapped in filter paper and then in a towel; it is left in the wrapped state for a time sufficient to ensure natural cooling. If a crystal is very sensitive even to small temperature drops, it should be allowed to cool in a thermostat. This is done by removing the cover of the crystallizer and introducing a rubber tube into the crystallizer through which the solution is removed. The cover is then replaced and the thermostat is switched off. If, in spite of all these precautions, a crystal cracks after its extraction, during storage or subsequent treatment, annealing is indicated. Annealing reduces the internal stresses which result from a nonuniform distribution of impurities captured during growth (heterometry). To anneal a crystal after the removal of the solution, the temperature in the thermostat is raised and the crystal is held in the empty crystallizer for a time which has to be determined experimentally. At the end of the annealing, the temperature is reduced gradually to room temperature and the crystal is removed. One must stress that such annealing alters the surfaces of a crystal very considerably.

A crystal should not be touched by hand but a crystal holder should be used to extract it from the crystallizer. This is particularly important in the case of crystals which dissolve easily. Contact with warm and moist fingers may crack a crystal or etch it. The surface of such a crystal may lose its luster and the fine details of the surface relief may be lost (such details are of considerable interest in studies of the growth of crystals). Fingerprints resulting from direct handling of a crystal can be removed only by dissolving the surface layer.

After it has been cooled and dried, a crystal should not be separated from its holder. This usually cracks the crystal. The separation can be carried out safely only if the end of the crystal holder is conical or if the seed is located in a cavity of a plate. Even in the case of a cylindrical rod it is not always possible to separate a crystal from its holder. However, in general, there is no need to separate a crystal from its holder. The holder rod may be cut at a distance of 3-4 cm from a crystal. The end of the rod projecting from the crystal can be used to hold the crystal during cutting.

If, during its extraction, a crystal passes through a layer of vaseline oil floating on the top of the solution, it should be cooled and then carefully washed twice in aviation-grade gasoline, and then dried rapidly with filter paper. Crystals of substances with a high vapor pressure (naphthalene, acetoxime) or those which become dehydrated in air (magnesium sulfate) should be stored in an inert liquid (for example, vaseline oil) or in a hermetically sealed container of volume not much larger than the volume of the crystal. Hygroscopic crystals which dissolve in air (potassium carbonate) should be stored in containers with tight covers smeared with Mendeleev grease; silica gel or calcium chloride should be placed in such containers to keep them dry. Crystals should be stored in such a way that they do not rub against one another. One must also remember that some crystals are photolyzed when exposed to light, or that they may react with paper, cotton wool, wood, or metals. Thus, crystals of iodic acid react with ordinary paper. It is difficult to give a comprehensive list of precautions necessary in the storage of various crystals; this section is simply intended to stress that great care should be exercised.

§30. Identification of Crystals

When crystals of well-known substances are grown from well-known solvents, usually there is no question as to the identity of the product. However, if the solute−solvent system has not been investigated thoroughly and if the phases formed at various temperatures are not known, it is necessary to check the identity of the crystals obtained.

In principle, a reagent may suffer three types of change during dissolution and subsequent crystallization. These changes are described below.

1. The number of molecules of the water of crystallization may change with the crystallization conditions. Thus, when $MgSO_4$ crystallizes between 1.8 and 48.1°C, $MgSO_4 \cdot 7H_2O$ crystals are obtained; up to 70°C, $MgSO_4 \cdot 6H_2O$ crystals are produced; at still higher temperatures, $MgSO_4 \cdot H_2O$ is grown. All these crystals have different structures, symmetries, and habit. Moreover, their physical properties are different.

2. Polymorphic transitions (changes in the structure without a change of chemical composition) may also take place. Here, we may cite ammonium nitrate, which can exist in four different polymorphic modifications with different structures in the temperature range from −16.9 to +100°C: polymorphic transitions alter the physical properties (including optical properties) as well as the habit of these crystals.

3. Finally, the dissolution or the reaction with impurities present in a solution may produce a new chemical compound. Here, we can mention the classical example of the effect of urea on the shape of ammonium chloride (NH_4Cl), which results in the formation of a new substance [Rimsky (1960)]. Therefore, one must always be aware of the possibility of reactions

and one should use some method to check the identity of the phase obtained by crystallization. Frequently it is sufficient to compare the shape of the crystals obtained with the illustration given in Groth's handbook (1906-1919). If the crystals obtained can also exist as minerals, one can use the old but very comprehensive mineralogical atlas of Goldschmidt (1913), the handbook of Dana et al. (1944-1962), Betekhtin's "Mineralogy" (1951) and also the handbook called "Minerals" (1960-1965). Extensive information on crystals is given in the "Barker Index of Crystals" (1951-1964). However, such a simple comparison is not necessarily reliable. A particular structure may appear under different crystallization conditions in a variety of forms (calcite and barite) or completely different substances may form crystals of apparently identical form (ammonium chloride and the urea$-NH_4Cl$ compound referred to earlier). Therefore, the goniometric method, which makes it possible to describe very accurately the external habit of a crystal, is not used nowadays to identify a substance.

Chemical analysis simply gives the composition of a crystal. Bearing in mind that it is difficult to distinguish chemically a substance containing seven water molecules from one which has six molecules of water of crystallization, it follows that various polymorphic modifications are completely indistinguishable by chemical means.

Much more reliable results can be obtained quickly, using several milligrams of a substance, by optical immersion studies [Tatarskii (1965)]. Information on the optical properties of synthetic inorganic compounds are given in books by A. N. and H. Winchell (1964) and Kordes (1960); for organic substances reference should be to A. N. Winchell's book (1954). The properties of naturally occurring compounds are given by A. N. and H. Winchell (1929-1951) and by Larsen and Berman (1934).

The most effective identification method is, of course, x-ray diffraction in which very small amounts of a substance can be used. This method not only gives the structure of the compound obtained but also indicates the presence of impurities whose nature can be found by spectroscopic analysis. X-ray diffraction analysis is a standard method in university and other research laboratories.

CONCLUSIONS

The art of growing crystals is gradually becoming a science. Considerably progress has been made in establishing a theoretical basis of crystal growth. Even fifteen years ago, we had formidable difficulties to overcome. Now, the collected observations and the various theories and hypotheses fit largely into a single logical system and a unified point of view. The technological side of the art of growing crystals has improved considerably. We have learned how to maintain stable conditions and partly how to control crystallization processes. Further development of crystal growth is retarded by the lack of standard commercially available equipment for the preparation of crystals from solutions and for investigations of crystal growth. The success of an experiment depends frequently on the failure-free operation of apparatus for many weeks and months, which is difficult to achieve using nonstandard equipment.

In many cases, we have learned how to prevent some undesirable features in the growth of crystals but much remains to be done.

The main difficulties in the synthesis of crystals are of a chemical nature: complex processes are necessary to purify a substance and to retain its purity; moreover, measurement of the purity involves laborious microanalysis and requires highly trained workers. The finer features of the interaction between impurities and growing crystals are not well known. Next to nothing is known about the structure of saturated and, particularly, supersaturated solutions or solutions containing impurities. These difficulties are, of course, common to the art of crystal growth and modern chemistry.

The authors do not wish the readers to regard this book as a set of mandatory procedures and directions. We have attempted to write this book so as to leave the reader free to choose his own growth method, construct his own apparatus, etc. There is full scope for initiative and creative thought in the growth of crystals. Much of what is now accepted practice may later be found to be incorrect.

LITERATURE CITED

Abdrakhmanova, N. F., and Deryagin, B. V., Dokl. Akad. Nauk SSSR, 120(1):94-97 (1958).

Akhmatov, A. S., Molecular Physics of Boundary Friction [in Russian], Fizmatgiz (1963).

Akhumov, E. I., Investigations of Supersaturated Aqueous Salt Solutions [in Russian], Goskhimizdat (1960).

Ansheles, O. M., Introduction to Crystallography [in Russian], Izd. LGU (1952).

Ansheles, O. M., Tatarskii, V. B., and Shternberg, A. A., Rapid Growth of Uniform Crystals from Solutions [in Russian], Leninizdat (1945).

Automatic Devices and Controllers (Handbook) [in Russian], Mashinostroenie (1964).

Balandin, A. A., Multiple Theory of Catalysis [in Russian], Parts 1-2, Izd. MGU (1963-4).

Barker Index of Crystals: Vol. 1, Part 1 (1951), ed. by M. W. Porter and R. C Spiller; Vol. 1, Part 2 (1951), ed. by M. W. Porter and R. C. Spiller; Vol. 2, Parts 1-3 (1956), ed. by M. W. Porter and R. C. Spiller; Vol. 3, Parts 1-2 (1964), ed. by M W. Porter and L. W. Codd; W. Heffer and Sons, Cambridge, England.

Belyustin, A. V., Kristallografiya, 6(5):807-808 (1961).

Belyustin, A. V., and Stepanova, N. S., Kristallografiya, 10(5):743-744 (1965).

Berg, L. G., Priroda, No. 7, pp. 95-97 (1957).

Berlin, A. Ya., Laboratory Techniques in Organic Chemistry [in Russian], Goskhimizdat (1963).

Betekhtin, A. G., Mineralogy [in Russian], Gosgeoltekhizdat (1951).

Bliznakov (Bliznakow), G., and Kirkova (Kirkowa), E., Z. Physik. Chem. (Leipzig), 206(3/4):271-280 (1957).

Borisov, V. T., Dukhin, A. I., and Matveev, Yu. E., Some Problems in the Theory of Growth of Crystals in Metallic Systems (Problems in Metallography and Physics of Metals) [in Russian], No. 8, Izd. "Metallurgiya," Moscow (1964), pp. 269-280.

Brauns, R., Chemical Mineralogy [Russian translation] (1904), pp. 259-261.

Buckley, H. E., Crystal Growth, J. Wiley, New York (1951).

Burton, W. K., Cabrera, N., and Frank, F. C., Phil. Mag., A243:299-358 (1951).

Byteva, I. M., Kristallografiya, 10(1):130-131 (1965).

Cabrera, N., and Vermilyea, D. A., "The growth of crystals from solution," in: Growth and Perfection of Crystals (Proc. Intern. Conf. on Crystal Growth, Cooperstown, New York, August 27-29, 1958), (ed. by R. H. Doremus, B. W. Roberts, and D. Turnbull), J. Wiley, New York (1958), pp. 393-410.

Cahn, J. W., Acta Met., 8(8):554-562 (1960).

Cahn, J W., Hillig, W. B., and Sears, G. W., Acta Met., 12(12):1421-1439 (1964).

Carlson, A., "The fluid mechanics of crystal growth from solution," in: Growth and Perfection of Crystals (Proc. Intern. Conf. on Crystal Growth, Cooperstown, New York, August 27-29, 1958), (ed. by R. H. Doremus, B. W. Roberts, and D. Turnbull), J. Wiley, New York (1958), pp. 421-426.

Cartier, R., Pindzola, D., and Bruins, P. F., Ind. Eng. Chem., 51(11):1409-1411 (1959).
Chemist's Handbook, Vol. 1 (1962), Vol. 2 (1963), Vol. 3 (1964), Vol. 4 (1965), Vol. 5 (1966), Goskhimizdat and Khimiya, Moscow.
Chernousov, N. P., Kutin, A. N., and Fedorov, V. F., Hermetically Sealed Chemical Engineering Machinery and Apparatus [in Russian], Mashinostroenie (1965).
Chernov, A. A., Usp. Fiz. Nauk, 73(2):277-332 (1961).
Chernov, A. A., "Some aspects of the theory on crystal growth forms in the presence of impurities," in: Adsorption et Croissance Cristalline, Paris (1965), pp. 265-277.
Chirvinskii, P. N., Preparation of Synthetic Minerals in Nineteenth Century [in Russian], Izd. Kievskogo Un-ta, Kiev (1904-1906).
Chistyakov, Yu. D., and Lainer, B. D., "Oriented overgrowth (epitaxis) of crystalline materials," in: Growth of Crystals (ed. by A. V. Shubnikov and N. N. Sheftal'), Vol. 4, Consultants Bureau, New York (1966), pp. 171-182.
Crystal Data: Determinative Tables, 7th ed., Washington (1963).

Dana, J. D., et al., The System of Mineralogy, 7th ed., Vols. 1-3, J. Wiley, New York (1944, 1951, 1962).
Danilov, V. I., Structure and Crystallization of Liquids [in Russian], Izd. AN UkrSSR, Kiev (1956).
Dankov, N. D., and Shishakov, N. A., Zh. Fiz. Khim., 23(9):1031-1035 (1949).
de Boer, J. H., The Dynamical Character of Adsorption, 1st ed., Clarendon Press, Oxford (1953).
de Boer, J. H., "Adsorption phenomena," in: Catalysis: Some Problems in the Theory and Technology of Organic Reactions [Russian translation], IL (1959).
de Haan, Y. M., Nature, 200(4909):876 (1963).
Dekeyser, W., and Amelinckx, S., Les Dislocations et la Croissance des Cristaux, Masson, Paris (1955).
Deryagin, B. V., Karasev, V. V., and Zorin, Z. K., "Special aggregate state of liquids in layers adjacent to solid surfaces," in: Structure and Physical Properties of Matter in the Liquid State [in Russian], Izd. Kievskogo Un-ta (1954), pp. 141-158.
Dislocations in Crystals: Bibliographical Guide [in Russian], Izd. AN SSSR (1960).
Dislocations in Crystals: Bibliographical Guide [in Russian], No. 2, Nauka (1966).
Doladugina, V. S., and Berezina, E. E., "Examination of corundum for homogeneity with a polarizing shadow system," in: Growth of Crystals (ed. by N. N. Sheftal'), Vol. 5B, Consultants Bureau, New York (1968a), pp. 177-186.
Doladugina, V. S., and Berezina, E. E., "An interference method of examining synthetic corundum crystals for homogeneity," in: Growth of Crystals (ed. by N. N. Sheftal'), Vol. 5B, Consultants Bureau, New York (1968b), pp. 187-192.
Dubrovo, S. K., Glass for Laboratory Instruments and Chemical Apparatus [in Russian], Nauka (1965).
Dukova, E. D., Kristallografiya, 5(5):813-815 (1960).

Encyclopedic Physics Dictionary [in Russian], Vols. 1-5, Sovetskaya Entsiklopediya (1962-1966).
Ermakov, N. P., Investigations of Mineral-Forming Solutions [in Russian], Izd. Khar'kovskogo Un-ta (1950).
Evzikova, N. Z., Zap. Vsesoyuzn. Miner. Ob-va, No. 94/2, pp. 129-147 (1965).

Feates, F. S., and Ives, D. J. G., J. Chem. Soc. 75:2798-2812 (1956).
Frank, F. C., New Investigations in Crystallography and Crystal Chemistry (Discussion of Part 1) [Russian translation], No. 1, IL (1950), p. 56.

Frank, F. C., Adv. Phys., 1(1):91-109 (1952).
Frank-Kamenetskii, V. A., Nature of Structural Impurities in Minerals [in Russian], Izd. LGU (1964).
Frenkel, J., J. Phys. USSR, 9(5):392-398 (1945).

Gavrilova, I. V., "Some aspects of the growth of crystals of sorbitol hexaacetate," in: Growth of Crystals (ed. by A. V. Shubnikov and N. N. Sheftal'), Vol. 3, Consultants Bureau, New York (1962), pp. 205-207.
Gavrilova, I. V., "Effects of crystallization temperature on the shape and homogeneity of crystals grown from aqueous solutions," in: Growth of Crystals (ed. by N. N. Sheftal'), Vol. 5A, Consultants Bureau, New York (1968), pp. 9-15.
Gavrilova, I. V., and Kuznetsova, L. I., "Aspects of the growth of monocrystals of potassium dihydrogen phosphate," in: Growth of Crystals (ed. A. V. Shubnikov and N. N. Sheftal'), Vol. 4, Consultants Bureau, New York (1966), pp. 69-72.
Gmelins Handbuch der anorganischen Chemie, 8th ed., Berlin (1924-1963).
Gol'danskii, V. I., and Chirkov, N. M., Dokl. Akad. Nauk SSSR, 58(6):1065-1067 (1947).
Gol'dshtein, Ya. R., Byull. Vses. Khim. Ob-va im. Mendeleeva, No. 10, pp. 15-17 (1941).
Goldschmidt, V., Atlas der Krystallformen, Heidelberg (1913).
Gorbachev, S. V., and Shlykov, A. V., Zh. Fiz. Khim., 29(5):797-801 (1955).
Griessbach, R., Austauschadsorption in Theorie und Praxis, Akademie-Verlag, Berlin (1957).
Groth, P., Chemische Krystallographie, Vols. 1-5, Leipzig (1906-1919).
Growth of Crystals, Vol. 1 (1958), Vol. 2 (1960), Vol. 3 (1962), Vol. 4 (1966), Vols 5A and 5B (1968), Vols. 6A and 6B (1968), Vol. 7 (1969), Vol. 8 (in preparation), Consultants Bureau, New York.

Handbook of Chemistry and Physics, Vols. 1-2, Cleveland, Ohio (1955).
Handbook on Fusibility of Systems of Anhydrous Inorganic Salts: Vol. 1, Binary Systems; Vol. 2, Ternary, Mixed Ternary, and More Complex Systems [in Russian], Izd. AN SSSR (1961).
Handbook on Solubility [in Russian], Vol. 1, Part 1 (1961); Vol. 1, Part 2 (1962); Vol. 2, Parts 1 and 2 (1963); Izd. AN SSSR, Moscow.
Handbook on Solubility of Salt Systems [in Russian], Vol. 1 (1953), Vol. 2 (1954), Vol. 3 (1961), Vol. 4 (1963), Goskhimizdat, Moscow.
Harmful Substances in Industry (ed. by N. V. Lazarev), Part 1, Organic Substances; Part 2, Inorganic Substances; [in Russian], Goskhimizdat (1963).
Hartmann, H., Monatsber. Deutsch. Akad. Wiss., Berlin, 6(2):108-114 (1964).
Henisch, H. K., Dennis, J., and Hanoka, J. J., J. Phys. Chem. Solids, 26(3):493 (1965).
Herrington, E. F. G., Zone Melting of Organic Compounds, Blackwells, Oxford (1963).
Hirth, J. P., and Pound, G. M., Condensation and Evaporation: Nucleation and Growth Kinetics (Progress in Materials Science, Vol. 11, ed. by B. Chalmers), Pergamon Press, Oxford (1963).
Holden, A., and Singer, P., Crystals and Crystal Growing, Doubleday, New York (1960).
Honigmann, B., Gleichgewichts- und Washstumformen von Kristallen, D. Steinkopf, Darmstadt (1958).
Houghton, J., Chem. Process. Eng., 46(12):639-646 (1965).

Ioffe, B. V., Refractometric Methods in Chemistry [in Russian], Goskhimizdat (1960).
Ion Exchange and Its Applications (Collection of Papers) [in Russian], Izd. AN SSSR (1959).

Jha, S. D., Kolloid. Z., 137(2/3):162-163 (1954).
Jona, F., and Shirane, G., Ferroelectric Crystals, Pergamon Press, Oxford (1962).

Kaishev, R., "Formation of nuclei and growth of layers on dislocation-free silver crystals," in: Abstracts of Seventh International Congress of Crystallographers [in Russian], Nauka (1966), p. 245.

Karpenko, A. G., et al., Kristallografiya, 6(1):146 (1961).

Kasatkin, A. P., Dokl. Akad. Nauk SSSR, 154(4):827-828 (1964).

Kasatkin, A. P., Kristallografiya, 10(4):550-554 (1965).

Kasatkin, A. P., Petrov, T. G., and Treivus, E. B., Kristallografiya, 6(6):952-954 (1962).

Kašpar, J., Barta, Č., and Nigrinova, J., "A new method for growing calcite without pressure," in: Growth of Crystals (ed. by N. N. Sheftal'), Vol. 6A, Consultants Bureau, New York (1968), pp. 5-6.

Keith, H. D., and Padden, F. J., J. Appl. Phys., 34(8):2409 (1963).

Khadzhi, V. E., Miner. Sb., No. 28/3, pp. 418-423 (1966).

Khaimov-Mal'kov, V. Ya., "The thermodynamics of crystallization pressure," in: Growth of Crystals (ed. by A. V. Shubnikov and N. N. Sheftal'), Vol. 2, Consultants Bureau, New York (1960), pp. 3-13.

Khaimov-Mal'kov, V. Ya., "Experimental measurement of crystallization pressure," in: Growth of Crystals (ed. by A. V. Shubnikov and N. N. Sheftal'), Vol. 2, Consultants Bureau, New York (1960), pp. 14-19.

Khaimov-Mal'kov, V. Ya., "The growth conditions of crystals in contact with large obstacles," in: Growth of Crystals (ed. by A. V. Shubnikov and N. N Sheftal'), Vol. 2, Consultants Bureau, New York (1960), pp. 20-29.

Kirillin, V. A., and Sheindlin, A. E., Thermodynamics of Solutions [in Russian], Gosenergoizdat (1956).

Kleber, W., Z. Physik. Chem. (Leipzig), 206(5/6):327-328 (1957).

Kondrat'ev, V. N., Structure of Atoms and Molecules [in Russian], Fizmatgiz (1959).

Kordes, E., Optische Daten., Weinheim (1960).

Kovalevskii, A. N., "A precision method of determining the saturation temperature of transport solutions," in: Growth of Crystals (ed. by A. V. Shubnikov and N. N. Sheftal'), Vol. 1, Consultants Bureau, New York (1958), pp. 264-266.

Kozlovskii, M. I., Kristallografiya, 3(4):483-487 (1958).

Krell, E., Handbuch der Laboratoriums-Destillation, M. Nijhoff, The Hague (1957).

K'uang Hang-Hang, K., Vestnik LGU, Ser. Geol. i Geogr., No. 6/1, pp. 146-149 (1966).

Kuznetsov, V. D., Crystals and Crystallization [in Russian], Gostekhteoretizdat (1953).

Landau, L. D., and Lifshitz, E. M., Course of Theoretical Physics: Vol. 5, Statistical Physics, 2nd ed., Pergamon Press, Oxford (1968).

Larsen, H. S., and Berman, H., "Microscopic determination of non-opaque minerals," U. S. Geol. Survey Bull. No. 848 (1934).

Lashchinskii, A. A., and Tolchinskii, A. R., Fundamentals of Construction and Design of Chemical Apparatus [in Russian], Mashgiz (1963).

Lazarev, N. V., Harmful Substances in Industry: Part 1, Organic Substances, Part 2, Inorganic Substances [in Russian], Goskhimizdat (1963).

Lemmlein, G. G., Sectorial Structure of Crystals [in Russian], Izd. AN SSSR (1948).

Lemmlein, G. G., and Dukova, E. D., Kristallografiya, 1(3):351-355 (1956).

Lengyel, S., "Investigation of partial solvent density in aqueous solutions of electrolytes," in: Thermodynamics and Structure of Solutions [in Russian], Izd. AN SSSR (1959), pp. 144-151.

Levich, V. G., Physicochemical Hydrodynamics [in Russian], Fizmatgiz (1959).

Lewin, S. Z., J. Phys. Chem., 59(10):1030-1034 (1955).

Lisgarten, N. D., and Blackman, M., Nature, 178(4523):39-40 (1956).

Lucas, H. J., Organic Chemistry, 2nd ed., Eurasia Publ. House (1962).

Mandelkern, L., Crystallization of Polymers, McGraw-Hill, New York (1964).
Mathieu, J. P., and Lounsbury, M., Compt. Rend., 229:1315 (1949).
Matusevich, L. N., Zh. Prikl. Khim., 32(3):536 (1959).
Matusevich, L. N., Zh. Prikl. Khim., 33(2):317 (1960).
Mellor, J. W., A Comprehensive Treatise on Inorganic and Theoretical Chemistry, London–New York–Toronto (1922-1964).
Mikhailov, I. G., and Syrnikov, Yu. P., Zh. Strukt. Khim., 1(1):12-27 (1960).
Minerals (Handbook) [in Russian], Izd. AN SSSR, Moscow: Vol. 1 (1960), Vol. 2, No. 1 (1963); Nauka, Moscow: Vol. 2, No. 2 (1965).
Mishchenko, K. P., "Thermodynamic properties of water in electrolyte solutions," in: Thermodynamics and Structure of Solutions [in Russian], Izd. AN SSSR (1959), pp. 97-105.
Mokievskii, V. A., Zap. Vsesoyuzn. Miner. Ob-va, No. 47/2, pp. 135-141 (1948).
Mokievskii, V. A., "Dependence of the rate of growth of magnesium sulfate crystals on the temperature and supersaturation of solutions," in: Crystallography [in Russian], Metallurgizdat (1951), pp. 239-244.
Morse, H. W., and Pierce, G. W., Z. Physik. Chem., 45(5):589-607 (1903).
Mullin, J. W., Crystallization, Butterworths, London (1961).

Neels, H., and Steinicke, U., Freiberger Forschungshefte, A267:433-442 (1963).
Nepomnyashchaya, V. N., Shternberg, A. A., and Gavrilova, I. V., "A laboratory method of growing large regular crystals and oriented blocks of lithium sulfate," in: Growth of Crystals (ed. by A. V. Shubnikov and N. N. Sheftal'), Vol. 3, Consultants Bureau, New York (1962), pp. 208-212.
New Technological Materials [in Russian], Khimiya (1964).

Ostroumov, G. A., Free Convection in an Enclosed Volume [in Russian], Gostekhteoretizdat (1952).

Palatnik, L. S., and Papirov, I. I., Directional Crystallization [in Russian], Metallurgiya (1964).
Palmer, L. S., Cunliffe, A., and Hough, J. M., Nature, 170:796 (1952).
Parvov, V. F., Kristallografiya, 9(4):584-585 (1964).
Peibst, H., and Noack, J., Z. Physik. Chem. (Leipzig), 221(1/2):115-120 (1962).
Petrov, T. G., Kristallografiya, 2(6):777-780 (1957).
Petrov, T. G., Method for Growing Single Crystals (Author's Certificate No. 136,057) [in Russian] (1960).
Petrov, T. G., Method for Growing Single Crystals of Given Shape from Solutions (Author's Certificate No. 151,299) [in Russian] (1962).
Petrov, T. G., Kristallografiya, 9(4):541-545 (1964).
Petrov, T. G., and Rozhnova, G. A., "Influence of impurities on the growth of crystals from aqueous solutions," in: Abstracts of Papers presented at the Third Conf. on Growth of Crystals [in Russian], Izd. AN SSSR (1963).
Petrov, T. G., and Treivus, E. B., Kristallografiya, 5(3):452-458 (1960).
Pfann, W. G., Zone Melting, 1st ed., J. Wiley, New York (1958).
Pis'mennyi, V. A., Zap. Vsesoyuzn. Miner. Ob-va, No. 89/6, pp. 699-704 (1960).
Polukarov, Yu. M., and Semenova, Z. V., Élektrokhimiya, 2(2):184 (1966).
Popov, G. M., and Shafranovskii, I. I., Crystallography [in Russian], Vysshaya Shkola (1964).
Popov, S. K., and Sheftal', N. N., Method of Dynamic Growth of Single Crystals (Author's Certificate No. 108,256) [in Russian] (1946).
Portnov, V. N., and Belyustin, A. V., Kristallografiya, 10(3):362-367 (1965).

Read, W. T., Jr., Dislocations in Crystals, McGraw-Hill, New York (1953).
Reviews on Crystallization (all published in Ind. Eng. Chem.):
  McCabe, W. L., 38(1):18-19 (1946);
  Grove, C. S., and Gray, J. B., 40(1):11-13 (1948);
  Grove, C. S., and Gray, J. B., 41:22-25 (1949);
  Grove, C. S., and Gray, J. B., 42:28-31 (1950);
  Grove, C. S., and Gray, J. B., 43:58-62 (1951);
  Grove, C. S., and Gray, J. B., 44:41-45 (1952);
  Grove, C. S., and Gray, J. B., 45:34-38 (1953);
  Grove, C. S., Schoen, H. M., and Palermo, J. A., 46:75 (1954);
  Palermo, J. A., Grove, C. S., and Schoen, H. M., 47(3, Part 2):520-523 (1955);
  Palermo, J. A., Grove, C. S., and Schoen, H. M., 48(3, Part 2):486-491 (1956);
  Palermo, J. A., Grove, C. S., and Schoen, H. M., 49:470-475 (1957);
  Schoen, H. M., and Grove, C. S., 50:430-434 (1958);
  Grove, C. S., and Schoen, H. M., 51:346-351 (1959);
  Schoen, H. M., 52(2):173-177 (1960);
  Schoen, H. M., 53(2):155-158 (1961);
  Schoen, H. M., and Van den Bogaerde, J. M., 54(4):57-62 (1962);
  Palermo, J. A., and Bennett, G. F., 56(10):38-52 (1964);
  Palermo, J. A., and Bennett, G. F., 57(11):68 (1965);
  Palermo, J. A., and Lin, K. H., 58(11):67 (1966).
Riddick, J. A., and Toops, E. E., Organic Solvents: Physical Properties and Methods (Vol. 7 of Technique of Organic Chemistry, ed. by A. Weissberger), 2nd ed., Interscience, New York (1955).
Rimsky, A., Bull. Soc. Franc. Mineral. Crist., 83(7-9):187-200 (1960).
Robinson, R. A., and Stokes, R. H., Electrolyte Solutions, 2nd ed., Butterworths, London (1959).
Rogacheva, É. D., and Belyustin, A. V., "Ratios of right- and left-handed forms of $MgSO_4 \cdot 7H_2O$ crystals grown from aqueous solutions," in: Growth of Crystals (ed. by N. N. Sheftal'), Vol. 5B, Consultants Bureau, New York (1968), pp. 38-40.

Safety Rules for Chemical Laboratories [in Russian], 2nd ed., Khimiya (1964).
Samoilov, O. Ya., Structure of Aqueous Electrolyte Solutions and the Hydration of Ions, Consultants Bureau, New York (1965).
Sampson, J. L., and di Pietro, M. A., Rev. Sci. Instr., 34(10):1150-1151 (1963).
Sears, G. W., J. Chem. Phys., 27(6):1308-1309 (1957).
Seidell, A., Solubilities of Organic Compounds, 3rd ed., Vol. 2, Van Nostrand, New York (1941).
Seidell, A., and Linke, W. F., Solubilities of Inorganic and Organic Compounds, Suppl. to the 3rd ed., Van Nostrand, New York (1952).
Seidell, A., and Linke, W. F., Solubilities of Inorganic and Metal-Organic Compounds, 4th ed., Vol. 1 (1958), Vol. 2 (1965), Van Nostrand, New York.
Shafranovskii, I. I., Lectures on Crystalline Morphology of Minerals [in Russian], Izd. L'vovskogo Un-ta (1960).
Shafranovskii, I. I., Mineral Crystals [in Russian], Part, 1, Izd. LGU (1957); Part 2, Gosgeoltekhizdat (1961).
Shakhparonov, M. I., Introduction to the Molecular Theory of Solutions [in Russian], Gostekhteorizdat (1956).
Sheftal', N. N., "Real crystal formation," in: Growth of Crystals (ed. by A. V. Shubnikov and N. N. Sheftal'), Vol. 1, Consultants Bureau, New York (1958), pp. 5-27.
Shlykov, A. V., and Gorbachev, S. V., Zh. Fiz. Khim., 29(4):607-614 (1955).
Shternberg, A. A., Uch. Zap. LGU, Ser. Geol-Pochv. Nauk, No. 65/13, pp. 56-59 (1944).
Shternberg, A. A., Crystals in Nature and in Technology [in Russian], Uchpedgiz (1961).
Shternberg, A. A., Kristallografiya, 7(1):114-120 (1962).

Shubnikov, A. V., How Crystals Grow [in Russian], Izd. AN SSSR (1935).
Šip, V., and Vaniček, V., "New items of equipment for the production of monocrystals," in: Growth of Crystals (ed. by A. V. Shubnikov and N. N. Sheftal'), Vol. 3, Consultants Bureau, New York (1962), pp. 191-195.
Slavnova, E. N., Inzh.-Fiz. Zh., No. 3, pp. 106-109 (1963).
Slavnova, E. N., Gordeeva, N. V., and Sitnik, T. K., Kristallografiya, 10(5):677 (1965).
Šmid. J., and Sommer, F., "Hydro- and thermomechanical conditions in a cylindrical vessel," in: Growth of Crystals (ed. by N. N. Sheftal'), Vol. 7, Consultants Bureau, New York (1969).
Sokolov, N. D., Usp. Fiz. Nauk, 57(2):205 (1955).
Spice, J. E., Chemical Binding and Structure, Pergamon Press, Oxford (1964).
Structure Reports, Vols. 7-21, Utrecht (1940-1958).
Strukturbericht, Vols. 1-6, Leipzig (1931-1941).
Sviridov, V. V., Photochemistry and Radiation Chemistry of Solid Inorganic Substances [in Russian], Minsk (1964).
Syrnikov, Yu. P., Compressibility of Electrolyte Solutions and Some Problems in the Theory of these Solutions (Author's Abstract of Dissertation) [in Russian] (1958).
Szent-Györgyi, A., Bioenergetics, Academic Press, New York (1957).

Tal'nikova, T. I., and Cherkasova, V. A., Reference Sources on Organic Chemistry [in Russian], Izd. LGU (1961).
Tatarskii, V. B., Crystal Optics and Immersion Analysis [in Russian], Nedra (1965).
Tiller, W. A., J. Appl. Phys., 29(4):611-618 (1958).
Titova, A. G., and Petrov, T. G., "Preparation of yttrium ferrite garnet single crystals under dynamic conditions," in: Abstracts of Papers presented at a Symposium on Processes of Synthesis and Growth of Semiconducting Crystals and Films [in Russian], Izd. Sib. Otd. AN SSSR, Novosibirsk (1965).
Tolansky, S., Surface Microphotography, Longmans, London (1960).
Torgesen, J. L., and Horton, A. T., J. Phys. Chem., 67(2):376-381 (1963).
Torgesen, J. L., Horton, A. T., and Saylor, C. P., J. Res. Natl. Bur. Std. (U.S.), 67c(1):25-32 (1963).
Treivus, E. B., and Petrov, T. G., Zap. Vsesoyuzn. Miner. Ob-va, Ser. 2, No. 93/2, pp. 197-203 (1964).
Treivus, E. B., Petrov, T. G., and Kamentsev, I. E., Kristallografiya, 10(3):380-383 (1965).

Urazovskii, S. S., Molecular Polymorphism [in Russian], Izd. AN UkrSSR, Kiev (1956).

Valyus, N. A., Raster Optics [in Russian], Gostekhizdat (1949).
van Bueren, H. G., Imperfections in Crystals, 2nd ed., North-Holland, Amsterdam (1961).
Van Hook, A., Crystallization: Theory and Practice, Reinhold, New York (1961).
Vedeneev, V. I., et al., Breaking Energies of Chemical Bonds [in Russian], Izd. AN SSSR (1962).
Veinik, A. I., Thermodynamics [in Russian], Minsk (1965).
Verkhovskii, V. N., Techniques and Methods of Chemical Experiments in Schools [in Russian], Vol. 1, Uchpedgiz (1959).
Verma, A. R., Crystal Growth and Dislocations, Butterworths, London (1953).
Voitsekhovskii, V. N., Zap. Vsesoyuzn. Miner. Ob-va, No. 92/5, p. 587 (1963).
Voskresenskii, P. I., Laboratory Techniques [in Russian], 6th ed., Khimiya (1964).

Walker, A. C., and Buehler, E., Ind. Eng. Chem., 42:1369-75 (1950).
Weissberger, A. (ed.), Technique of Organic Chemistry, Vol. 7, Organic Solvents: Physical Properties and Methods of Purification (by J. A. Riddick and E. E. Toops), 2nd ed., Interscience, New York (1955).

LITERATURE CITED

Wells, A. F., Phil. Mag., 37(266):184-189 (1946); 37(267):217-236 (1946).

Wilke, K. T., Methoden der Kristallzüchtung, Berlin (1963).

Winchell, A. N., The Optical Properties of Organic Compounds, 2nd ed., Academic Press, New York (1954).

Winchell, A. N., and Winchell, H., Elements of Optical Mineralogy, Part 1, 5th ed. (1937); Part 2, 4th ed. (1951); Part 3, 2nd ed. (1931); J. Wiley, New York.

Winchell, A. N., and Winchell, H., The Microscopical Characters of Artificial Inorganic Solid Substances: Optical Properties of Artificial Minerals, 3rd ed., Academic Press, New York (1964).

Yur'ev, Yu. K., Practical Work on Organic Chemistry [in Russian], Izd. MGU (1964).

Zakhar'evskii, M. S., Kinetics and Catalysis [in Russian], Izd. LGU (1963).

Zdanovskii, A. B., Kinetics of Sólution of Natural Salts under Forced Convection Conditions [in Russian], Goskhimizdat (1956).